40th Biennial Meeting of the American Society of Sugar Beet Technologists 2019

AA006236

Anaheim, California, USA
27 February – 2 March 2019

ISBN: 978-1-7138-1370-5

Printed from e-media with permission by:

Curran Associates, Inc.
57 Morehouse Lane
Red Hook, NY 12571

Some format issues inherent in the e-media version may also appear in this print version.

Printed with permission by Curran Associates, Inc. (2020)

For permission requests, please contact the American Society of Sugar Beet Technologists at the address below.

American Society of Sugar Beet Technologists
800 Grant Street, Suite 300
Denver, Colorado 80203-2944

Phone: 303-832-4460
Fax: 303-832-4468

www.bsdf-assbt.org

Additional copies of this publication are available from:

Curran Associates, Inc.
57 Morehouse Lane
Red Hook, NY 12571 USA
Phone: 845-758-0400
Fax: 845-758-2634
Email: curran@proceedings.com
Web: www.proceedings.com

TABLE OF CONTENTS

GROULX, BRIAN J.[1], JAMES F. STEWART[1], COREY GUZA[2] AND DENNIS BISCHER[2].
[1]Michigan Sugar Company, Agricultural Research Center, 1459 S. Valley Center Dr., Bay City, MI 48706, [2]Michigan Sugar Company, 122 Uptown Drive, Suite 301, Bay City, MI 48708.
Managing *Rhizoctonia solani* root rot in sugarbeets in Michigan with registered and experimental fungicide applications.

EXPANDED ABSTRACT

Michigan sugarbeet growers lose an estimated 1 to 4 tons per acre to Rhizoctonia root rot infestations depending upon the region they farm in. Yield losses can be much higher when weather conditions prevent timely fungicide applications, when sprayer malfunctions occur or when the wrong variety is planted. Michigan Sugar Company has three distinct Rhizoctonia risk zones. The western region has a high risk of Rhizoctonia root rot and tolerant varieties in addition to Quadris applications are required to grow a successful crop. The central region has a moderate risk and the eastern growing region has a low to moderate risk of Rhizoctonia.

Rhizoctonia root and crown rot is caused by the fungus *Rhizoctonia solani*. This fungus is divided into several anastomosis groups (AGs). *R. solani* AG 2-2 IV and IIIB are the AGs that cause root and crown rot in sugarbeets. Surveys conducted by Dr. Linda Hanson, USDA-ARS, East Lansing Michigan, show that the IIIB strain is more virulent and prevalent than the IV strain.

Almost all of the Michigan sugarbeet growers utilize Quadris (azoxystrobin) in some manner. Approximately 50 percent of the growers apply Quadris in-furrow at planting followed by an 8 leaf foliar application directed at the sugarbeet crown. About 40 percent of the growers apply only a foliar Quadris application and about 10 percent apply only an in-furrow application.

Objective

The objective of this study was to compare the efficacy of registered and experimental fungicides for the control of *Rhizoctonia solani* under natural conditions in area with high disease potential. Nearly all of the Michigan Sugar Company growing region can benefit financially from including at least one fungicide application for the control of *Rhizoctonia solani*. Approximately 1/4 – 1/3 of the growing region would benefit from making two fungicide applications. Lastly, the worst areas for disease require growing the most tolerant varieties as well as multiple fungicide applications.

Method

The data used was generated from three trials performed in 2015, 2016, and 2017. The trials were small plot randomized complete block design. Trials ranged from four to six replications. Plots were planted in six 22 inch rows with the plot length being 38 feet. The four center rows were counted, rated, and taken to harvest for yield and sugar data. In-furrow applications were made in a 3.5 inch band over the row. The in-furrow application placement was after seed drop, but before closing the seed trench. Foliar applications were made at the 4, 8, and 12 leaf stage in a 7 inch band comparing those timings against each other. The goal was to target growing areas with consistently high levels of natural infection. Because of the possibility of the fungus developing resistance to azoxystrobin, several fungicides with different modes of action were evaluated by Michigan Sugar Company. Moncut (flutolanil) Group 7, Proline (prothioconazole) Group 3, Headline (pyraclostrobin) Group 11 and Xanthion (pyraclostrobin plus a bacillus species) Group 11 and 44 have demonstrated effective Rhizoctonia control in research trials.

Results

Quadris applied in-furrow (10 fl oz/A) at planting in a 3.5 inch T-band followed by a Quadris application at the 8 leaf stage (14 fl oz/A) in a 7 inch band has provided the best Rhizoctonia root rot control in Michigan as well as providing the highest yield and sugar content. When comparing the number of dead beets per 100 feet of row, in-furrow applications of Quadris at the 10 fl oz/A as well as the 14 fl oz/A also provided control that was not statistically different from two applications of Quadris. Proline SC applied at 5.7 fl oz/A in-furrow also gave comparable control. All Quadris in-furrow treatments reduced emergence significantly when compared to untreated plots, about 5% on average. This reduction in emergence was compensated by providing better disease control and achieving a higher harvestable stand. A one-time IF application of Quadris at the 10 oz. or 14.25 oz. rate provided significantly better results than a one-time IF application at 7 oz. rate, or a one-time foliar application at any rate or timing. Other products such as Proline SC, Xanthion, Priaxor, and Moncut SC applied either in furrow or foliar also provided control when compared to the untreated check. When only making foliar applications, the 8 leaf timing has been superior to either the 4 leaf timing or 12 leaf timing. Very late or "rescue" applications of Quadris at or near row closure have also been attempted with little or no success.

Conclusion

Quadris applications at both the in-furrow and 8 leaf stage are providing the best control of *Rhizoctonia solani* as well as providing the most economic return. A single application of Quadris at the in-furrow timing should be at a rate of at least 10 oz. per acre in 22" rows. A single in-furrow application of at least 10 oz. per acre provides superior control over a foliar application at any rate or timing tested. Other products tested provided control when compared to the untreated check and could be used for resistance management. Seed treatments for Rhizoctonia control have also been tested, but do not improve control over a program with two applications of Quadris. Michigan Sugar Company recommends that growers who are not currently applying Quadris in-furrow to be sure to use seed treated with one of the available seed treatments. All seed suppliers for Michigan offer a Rhizoctonia seed treatment as a standard or option on each unit of seed provided.

Evaluation of azoxystrobin alternatives and *Rhizoctonia solani* sensitivity screening in Michigan sugar beets

Jaime F. Willbur[1], Chris Bloomingdale[1], Douglas H. Minier[1], Linda E. Hanson[2]

[1]Michigan State University, Department of Plant, Soil and Microbial Sciences, East Lansing, MI;[2]United States Department of Agriculture – Agricultural Research Service, Department of Plant, Soil and Microbial Sciences, East Lansing, MI

Rhizoctonia root and crown rot (RRCR) is caused by *R. solani* Kühn and continues to be a major pest of sugar beets. In Michigan, Quadris (azoxystrobin) is widely applied to manage RRCR. Azoxystrobin, a quinone outside inhibitor (QoI), targets a single site to inhibit fungal respiration and so possesses a high risk of fungicide resistance development. Continued reliance on one to two applications of this product per season has prompted the evaluation of efficacious alternatives for RRCR management and preliminary fungicide sensitivity assays. In 2018, in-furrow and banded applications (made at the 6-8 leaf stage) of Serenade (*Bacillus subtilis* QST 713), Proline (prothioconazole), and Propulse (fluopyram+prothioconazole) were evaluated in an inoculated trial at the Saginaw Valley Research and Extension Center in Frankenmuth, MI. Stand counts were collected throughout the season and total stand loss recorded. At harvest, 20 arbitrarily selected beets were rated and assigned a severity score (0-7 scale). The Rhizoctonia root rot index (DIX) was calculated by multiplying the incidence (0-100%) by the mean symptomatic root severity divided by seven. A generalized linear mixed model procedure (SAS v. 9.4) was used to conduct the ANOVA (α=0.05). Mean comparisons were determined using Fisher's Protected LSD (α=0.05).

A complementary assay of azoxystrobin sensitivity was also conducted to determine the status of Quadris efficacy in Michigan *R. solani* populations. In 2018, *R. solani* isolates were collected from five commercial fields treated with azoxystrobin in the Michigan sugar beet growing region. Two additional baseline isolates, collected prior to azoxystrobin use in sugar beet, were included for comparison. Isolates were screened on half-strength potato dextrose agar (PDA) amended with salicylhydroxamic acid (SHAM) at 10 $\mu g\ ml^{-1}$ and azoxystrobin at concentrations: 0, 0.001, 0.01, 0.1, 1, and 10 $\mu g\ ml^{-1}$. After 72 hours, radial growth was measured along two transects. Percent inhibition was calculated using the equation: 100 - [(response amended / response non-amended)*100]. Significant radial growth was observed at highest concentrations, however, colony density was visibly inhibited. Methods were altered to measure colony mass in half-strength clarified V8 broth amended with SHAM at 10 $\mu g\ ml^{-1}$ and azoxystrobin at 0, 0.01, 0.1, 1, 10, and 100 $\mu g\ ml^{-1}$. Effective concentrations for 50% inhibition of colony mass (EC_{50}) were determined using three-parameter logistic regression (SAS v. 9.4).

Field treatments significantly affected total stand loss (P<0.0001), DIX (P<0.01), and yield (P<0.05). Only treatments of Serenade, Proline, and Propulse used in a program with Quadris yielded significantly higher than the non-treated control. Notably, Serenade+Quadris in-furrow followed by a banded application of Proline significantly reduced root rot scores (-30%) and resulted in higher yields (+7 T A^{-1}; 15692 kg ha^{-1}) than the Quadris only program. No products tested individually performed significantly differently than the Quadris program. In the amended broth studies, mean EC_{50} values were 0.013 and 0.018 $\mu g\ ml^{-1}$ for baseline and non-baseline isolates, respectively. Furthermore, isolates were sensitive to azoxystrobin concentrations below label rates (1.35-2.7 $\mu g\ ml^{-1}$). In this preliminary screen, azoxystrobin insensitivity was not observed in Michigan *R. solani* populations, however, additional testing is necessary to better represent the region.

Prune Pegot-Espagnet, Bruno Desprez, Brigitte Devaux, Pierre Devaux, Karine Henry, Nicolas Henry, Glenda Willems, Ellen Goudemand, Brigitte Mangin

ORIGINAL GENETIC DIVERSITY DISCOVERY COMPARING A SUGAR BEET ELITE PANEL WITH A PROGENY FROM (SUGAR BEET ELITE X EXOTIC) CROSS

ABSTRACT

Selection of stable varieties producing the highest amount of extractable sugar per hectare (ha), resistant to diseases and respecting environmental criteria is undoubtely the main target for sugar beet breeding. Introgressing exotic germplasm into breeding programmes can be a way to bring new interesting allelic diversity in breeding material.

Two sugar beet populations, an elite panel (2101 genotypes, phenotyped in 21 environments, genotyped with 629 SNPs) and (elite x exotic) progeny (187 genotypes, nine environments, genotyped with 303 SNPs), were compared for three traits: potassium quantity (K meq/100g), sodium quantity (Na meq/100g) and N-alpha-amino quantity (N meq/100g).

Genome-wide association studies (GWAS) were performed for each trait in each population with the multi-locus mixed model approach (MLMM) proposed by Segura et al. (2012). The eBIC parsimony criterion (Chen and Chen, 2008) was used to choose the MLMM model that explains best trait variability with the least SNPs. Then SNPs were merged into QTLs if they were located on the same chromosome and have linkage desequilibrium greater than a significance threshold, calculated for each population from corrected r2 (Mangin et al, 2012).

The number of selected SNPs was higher in the elite panel due to higher power of detection and larger diversity. Very close SNPs in both populations were found, e.g. a probably pleiotropic QTL for Na and N traits located at the middle of the chromosome 5. Differences in genetic architectures were also observed, as the SNP detected for Na in the second half of the 5th chromosome in the (elite x exotic) progeny for which there is not colocalized detected SNP in the elite panel for this trait. Moreover, some SNPs detected in the (elite x exotic) progeny showed a favorable effect of the exotic (i.e., the impurity quantity decreases with the presence of the exotic allele).

These results illustrate that the introgression of exotic alleles into elite germplasm is an interesting method to bring new useful variability in breeding material.

This Research program is funding by the French Government, under the management of the Research National Agency (ANR). AKER is included in the «Investissements d'avenir » as refered as ANR-11-BTBR-0007

References

Chen, J. and Chen, Z. Extended bayesian information criteria for model selection with large model spaces. Biometrika, 95(3):759–771, 2008.

B Mangin, A Siberchicot, S Nicolas, A Doligez, P This, and C Cierco-Ayrolles. Novel measures of linkage disequilibrium that correct the bias due to population structure and relatedness. Heredity, 108(3):285–291, March 2012. ISSN 0018-067X. doi: 10.1038/hdy.2011.73. URL https://www.ncbi.nlm.nih.gov/pmc/articles/PMC3282397/.

Segura, V., Vilhjálmsson, B. J., Platt, A., Korte, A., Seren, Ü., Long, Q., and Nordborg, M. An efficient multi-locus mixed-model approach for genome-wide association studies in structured populations. Nature genetics, 44(7):825–830, 2012.

AKER a French Research Initiative for a competitive Innovation for Sugar Beet

Christian Huyghe[1], Vincent Laudinat[2], Karine Henry[3], Fabienne Maupas[2], Bruno F. Desprez[3]*. [1]Institut National de la Recherche Agronomique, 147, Rue de l'Université, F-75338 Paris, France , [2]ITB, 45, rue de Naples , F-75008 Paris, France, and [3]Florimond Desprez Veuve & Fils SAS, BP41, 3, Rue Florimond Desprez, F-59242 Cappelle-en-Pévèle, France.

AKER is an 8 year programme (2012 to 2020) with 11 partners and a budget of around 20 Mio Euros with 5 Mio granted by the French Ministry of Research. As starting this initiative, all partners were convinced that breeding should play an important role in the capacity of sugarbeet to be competitive in front of cane. The challenge was to increase the rate of sugar yield gain (around 2% actually), and, as a more global challenge, creating varieties to meet high quality, safety and sustainable agriculture.

Global diversity available and its management to answer these questions is critical. Considering the funnel effect of breeding, AKER's project aims to widen this funnel maximizing the availability by exploring the genetic resources variability, and to shorten the time duration to the final elite production.

By taking into account all germplasm available including 50 world genebanks, and after genotyping, we estimated that a core collection of only 15 plants was sufficient to carry all allelic variability which is not encountered already in the elite germplasm.

After eight years of backcrossings with elite lines, by choosing the right complementary ones, we got 3,000 plants representing all variability coming from these 15 exotic plants into an elite background. The two years left within the program is to phenotype all this final core collection. We so expect to get a value of each of these small source of genomic variations. Bringing useful traits or dirty ones.

Another idea was to transferred this revolutionary methodology which we call genomic selection coming from cattle breeding to crop breeding, and particularly to sugar beet breeding. The goal is the increase the speed of genetic progress. We can summarize by acting that GS is really a promising tool, also for sugar beet breeding.

Concerning phenotyping, AKER is devoting 60% of its total budget. Remembering that we have to phenotype as much traits as possible the 3,000 hybrids from the Core Collection. We need high-throughput, non destructive, accurate and dynamic phenotyping. Phenotyping from seed, early growth stages to seedling and of course growing stage to harvest including diseases or any abiotic stress. Example such as seed phenotyping thanks to a three dimensional X-ray machine is shown. IR thermography to evaluate sugar content in young plantlets, MRI to get access to fat content of the embryo within dry seed, Hyper spectral image test (visible + IRP) to get access to biochemical content of organs. For the 2018 campaign, the field experiment consist of more than 70,000 plots specifically dedicated to AKER : to examine the 3,000 hybrids and some of the other AKER breeding experiment.

This Research program is funding by the French Government, under the management of the Research National Agency (ANR). AKER is included in the «Investissements d'avenir » as refered as ANR-11-BTBR-0007

ALKALIZATION PRETREATMENT OF THIN JUICE SUPPLIED TO WEAK ACID CATION (WAC) THIN JUICE SOFTENING

RHOTEN, CHRISTOPHER D., ESCON America, PO Box 1308, Black Mountain, NC 28711
Dr. Bernd-Christoph Schulze, ESCON GmbH, Schlossstrasse 48A, 12165 Berlin/Germany

Introduction

Operation of Weak Acid Cation (WAC) thin juice softening requires that the concentration of limesalts in hard thin juice supplied to the WAC ion exchange cells be less than 0.100g CaO/100dsS. During beet storage the organic acid concentration in beets increases resulting in limesalts concentration in hard thin juice often exceeding 0.100g CaO/100ds. When limesalts concentration in filtered hard thin juice increases above this operational limit, the juice alkalinity must be increased by addition of an alkalizing agent to correct the alkalinity deficiency thus lowering the limesalts concentration in hard thin juice supplied to the WAC ion exchange cells.

Soda ash is generally used for the purpose of juice alkalization. Proper operation of 2nd carbonation and correct soda ash addition reduces the limesalts concentration in hard juice to below 0.100g CaO/100ds usually approximating that in thin juice produced from fresh beet. Good control of the re-alkalization process results in thin juice having sufficiently low limesalts concentration for supply to the WAC softening cells while also having stable pH and color behavior during juice concentration and sugar end operations. When so supplied, the WAC softening system routinely produces soft thin juice having a suitably low limesalts concentration.

Relatively precise control of 2nd carbonation alkalinity and the subsequent addition of alkalizing agent, especially during periods of fluctuating beet quality, is essential to avoiding both under and over alkalization of the juice. Under alkalization results in limesalts concentration exceeding the operational limit to the WAC cells. Sustained and/or repeated excess limesalts concentration in hard thin juice supplied to the WAC cells leads to resin overloading and a reduction in effective resin capacity along with difficulty in fully restoring resin capacity during routine regeneration cycles. Over-alkalization, in spite of yielding sufficiently low limesalts concentration to the WAC cells, leads to high juice pH and increased color rise during juice concentration.

Quality Requirements for Hard Thin Juice Supply to WAC Softening Cells

The juice supplied to the WAC softening cells must meet certain quality requirements in order to insure optimal performance of the softening system. Certain characteristics of the hard juice supply influence both the capacity and life of the resin as well as the hydraulic efficiency of the resin beds in the WAC cells. The following hard juice characteristics are essential to optimal operation of a WAC softening system.

1. <u>The MAXIMUM juice hardness supplied to the WAC cells should not exceed 0.100g CaO/100ds</u>. Variation in limesalts concentration in hard thin juice is expected in normal factory operation. Due to such "normal" variation, the targeted average limesalts concentration in hard juice produced must be such that the limesalts concentration limit is never exceeded. For example, if

the standard deviation of limesalts in the hard juice produced is +/- 0.020g CaO/100DS, the average limesalts concentration in hard juice supplied to the WAC cells should not exceed about 0.064g CaO/100ds.

2. Post precipitation of calcium carbonate in filtered hard juice must be avoided or at least minimized to the extent possible. Juice exiting the 2nd carbonation gassing vessel is normally supersaturated with calcium carbonate. The 2nd carbonation post reaction tank serves to de-supersaturate the juice by fully precipitating the excess $CaCO_3$ in solution. Sufficient retention time and adequate agitation (or recirculation) are required to fully de-supersaturate the 2nd carbonation effluent. The post reaction tank is also an effective place for the final addition of soda ash for limesalts and juice pH control.
3. The hard juice supplied to the WAC cells must be essentially sediment free. It is essential that precipitated calcium carbonate is never allowed to pass to the WAC cells. If precipitated calcium carbonate accumulates in the WAC resin bed, the hydraulic capacity and softening efficiency of the resin bed is compromised due to channeling. The resin itself is also affected even to the point of irreversible fouling leading to premature resin replacement. It is essential that 2nd carbonation filtration be operated in such manner as to avoid even the occasional passage of calcium carbonate sediments. Double filtration is often practiced to avoid such excursions altogether.

Operation of Alkalization Pretreatment Process

When limesalts concentration in hard thin juice supplied to the WAC cells approaches the operational limit of 0.100g CaO/100ds, the alkali content of the juice must be increased to lower the concentration of soluble limesalts in the juice. While an alkalizing agent may be added anywhere in the beet end upstream of 2nd carbonation, it is most commonly added as soda ash (Na_2CO_3) either directly to 2nd carbonation or to the 2nd carb post reaction tank. The most precise control of juice alkalization may be achieved through a 2-step operation. The first step is the close control of the 2nd carbonation gassing at the "optimum" juice alkalinity in the absence of alkali addition. The second step is the relatively precise addition of soda ash to the 2nd carbonation post reaction tank to achieve an optimal alkali balance in the juice.

2nd Carbonation Operation

Second carbonation should be operated at or near the optimum alkalinity of the under-alkalized juice in process. This is essential to insure the lowest limesalts concentration in juice exiting the 2nd carbonation gassing vessel. It is quite important, relative to both lime salts control and proper overall juice alkalization, to maintain 2nd carbonation alkalinity near the "optimum" to enable minimum addition of alkali (soda ash) to achieve acceptable lime salts to the WAC units WITHOUT over-alkalizing the juice. Juice over-alkalization leads to pH increase during juice concentration, higher thick juice color and high pH values (>9.0 pH) in the sugar end.

It is relatively common in some areas of the USA for factories to process beets having relatively low natural alkalinity. Many of these factories do not normally employ any form of juice alkalization in the absence of juice softening. In such factories, 2nd carbonation is often operated at an alkalinity (pH) significantly higher than the optimum so as to provide a relatively higher thin juice pH to help support evaporator and sugar end pH profiles. When WAC juice softening is installed in a factory where soda addition has not been used to control juice pH stability, it is often a bit uncomfortable for operators to accept the operation of 2nd carbonation at the lower optimum alkalinity until it is demonstrated that the appropriate addition of an alkalizing agent to the juice satisfactorily compensates for the lower target alkalinity (pH) in 2nd carbonation.

Soda Ash Addition and Control

Soda ash may be introduced to the process as an alkalizing agent in a number of different locations in the beet end. As noted earlier, an optimal point of addition is to the 2nd carbonation post reaction tank for, at least, a final "trimming" addition of soda ash to that vessel. This approach may assure minimum soda ash usage and the opportunity for relatively precise control of the overall alkalization process.

Soda ash may be introduced directly to the juice stream as a solid in granular form or as a solution of soda ash at a standardized concentration. Improved control of juice alkalization is usually achieved by dissolving granular soda ash in condensate or soft thin juice to yield a 10-20% solution of soda ash by weight. The stock soda ash solution is then metered to the process in a controlled manner via a simple control valve or VFD equipped treatment pump. As the limesalts concentration in the 2nd carbonation effluent varies, the concentration of the soda ash stock solution may be adjusted relative to the demand for soda ash solution flow in order to remain within the control range of the soda ash dosing system.

The most common alkalization control approach is to manage the flow of stock soda ash solution to the process based on the flow of juice through the system. Normally, the flow of juice through 2nd carbonation filtration is selected for the flow control input. The flow of soda ash stock solution to the process is managed through a juice flow/ratio controller. As process juice flow varies, the flow of soda ash solution varies relative to process juice flow as a function of a "manually selected" flow ratio. When there is a change in juice quality relative to limesalts concentration, this flow ratio must be manually adjusted to maintain the correct limesalts concentration/juice alkalization targets for the process. Such a control arrangement depends on the observation of a change in 2nd carbonation or hard juice effluent limesalts concentration followed by an adjustment of the soda ash flow ratio to correct any deviation in the targeted hard juice limesalts.

This system of control is satisfactory as long as the juice quality is relatively stable and only occasional adjustment of the flow ratio is necessary to maintain the targeted hard juice limesalts concentration. However, when large routine variations in juice quality occur, such as are encountered with longer beet storage or when beets are received from different areas, the "manual" flow ratio control of juice alkalization often results in a series of over and under alkalized conditions. This is primarily due to the lag in adjustment of the "manual" flow ratio input relative to the juice in process as a result of the inherent 1) lag in laboratory data output relative to process condition, 2) process retention time relative to process control point and 3) any delay and/or error in corrective process adjustment.

This control lag may be largely eliminated with the "two-step" alkalization process described below. The first step is the control of 2nd carbonation at optimum alkalinity <u>without</u> soda ash addition FOLLOWED by the second step of soda ash addition to the juice exiting the 2nd carbonation gassing tank. The process may be automated to control the addition of soda ash to the 2nd carbonation post reaction tank.

This automatic control approach is relatively simple and depends upon operating the 2nd carbonation gassing at the optimum alkalinity without soda addition. In juice having higher organic acid content, the operation of 2nd carbonation at the optimum alkalinity in the absence of added soda ash will produce a juice having a relatively lower pH and the lowest limesalts concentration in the un-alkalized juice. Soda ash addition may then be controlled at a higher pH relative to the end point for optimally alkalized juice. The pH control set point generally approximates that of juice produced from freshly harvested beet. For such operating conditions and depending on the beet in process, the optimum pH of 2nd carbonation without soda ash addition is often in the range of 8.7-9.0 whereas the pH after correct soda addition is usually in the range of 9.2 to 9.5 pH. An example of this system of control is shown in Figure 1.

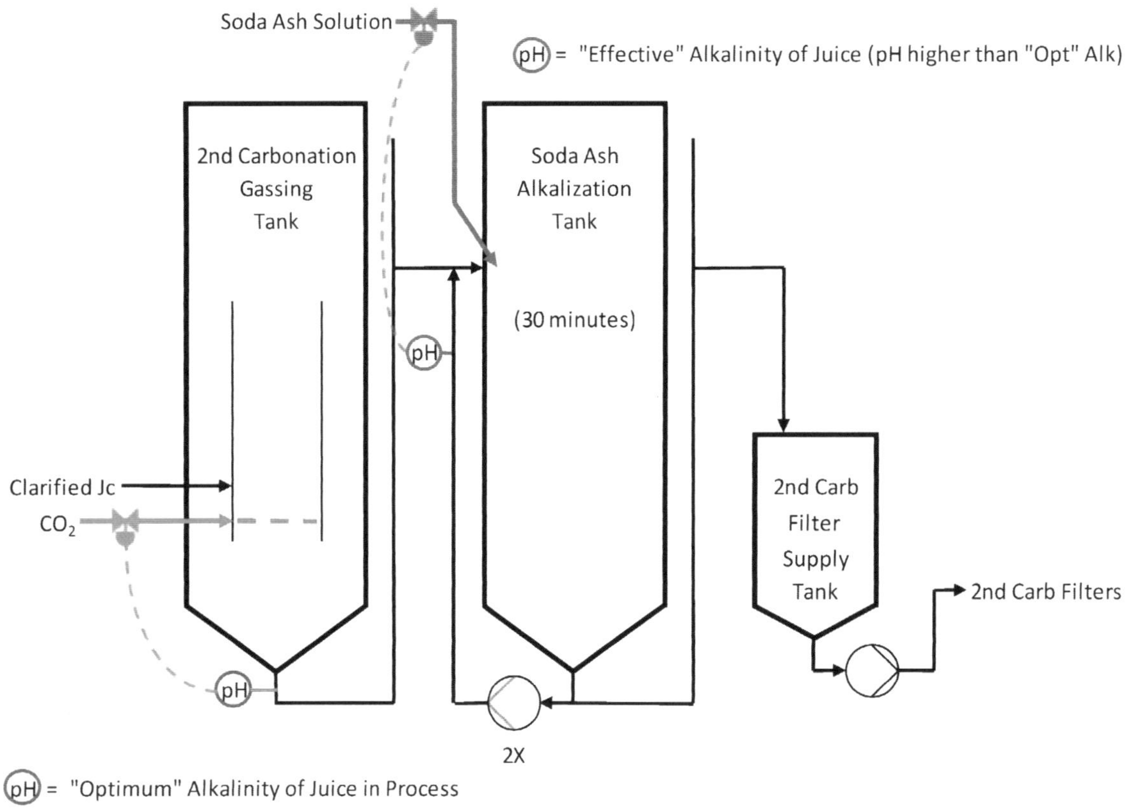

Double Alkalization Control of Total Juice Alkalization

Figure 1

The following points describe the important aspects in the control of the alkalization process using the "Double Alkalization" approach.

1. Double Alkalization gives acceptably consistent hard juice limesalts to the WAC cells while also providing optimal pH balance during juice concentration. Ideally, there should be no pH rise (due to juice over-alkalization) in the juice evaporator. A slight drop in pH of about 0.5 pH during juice concentration is quite common and usually acceptable. When over-alkalization occurs due to excess soda ash addition, pH and color rise across the evaporator increase relative to the degree of excess alkali in the juice.

2. As the alkalinity of juice in process declines due to beet deterioration, the difference between the "optimum" alkalinity (pH) of 2nd Carbonation and the "effective" alkalinity (pH) of optimally alkalized juice generally increases relative to an increase in the organic acid content of the juice in process.
3. To be most effective, 2nd Carbonation gassing must be operated as close to the optimum alkalinity as possible without entering the bicarbonate range and in the absence of added or excess alkalizing agent allowing a subsequent final alkali addition to the juice to be managed at minimum total soda ash usage. The operator need only make occasional small trimming adjustments to the pH set point of the "Soda Ash Alkalization" tank to manage the WAC supply hardness and optimal thick juice pH.

Benefits of Optimal Juice Alkalization

While the control of limesalts concentration to the WAC softening cells is the primary goal of juice alkalization pretreatment, it is only one aspect of the result of correct juice alkalization. As beet quality deteriorates due to storage or any other adverse condition the organic acid content of the beet increases. The addition of an alkalizing agent to the juice produced must vary with respect to the resulting change in alkalinity of the juice related to the degree of beet deterioration.

When limesalts concentration in the hard juice produced from 2nd carbonation increases above 0.100g CaO/100ds or begins to vary significantly, one might conclude that a simple solution is to simply increase soda ash addition to a point where the hard thin juice never exceeds this limit. While such a practice may largely avoid the problem of hard juice limesalts exceeding the operational limit to WAC softening, it largely ignores the potential consequences to thick juice quality.

When thin juice is over-alkalized, there will be an associated pH and color increase across juice concentration relative to the degree of over-alkalization. The color increase in the juice is directly proportional to the increase in pH. The greater the degree of over-alkalization the larger the pH and color increase. Figure 2 shows the relationship between thick juice pH and color rise across juice concentration in one North American factory. Similar curves have been observed for a number of different factories indicating minimum color rise during juice concentration in the range of 8.5 +/-0.3 pH in thick juice.

The example in Figure 2 is during relatively "normal" variation in beet quality in a factory operating a 5-effect Roberts evaporator. This factory must routinely vary the addition of soda ash to control lime salts to the WAC units. When juice alkalization is on target, hard thin juice limesalts concentration is well within operating limits to the WAC cells with thick juice pH in the range of 8.0-8.5 accompanied by the lowest color rise during juice concentration. During the period of lowest color rise thick juice averaged 8.1 pH with color rise of 620 IU. The hard juice limesalts during the same period averaged 0.075g CaO/100ds.

It must be added that, due to variation in beet quality during this same operating period, hard juice was under-alkalized 18% of the time. This resulted in WAC supply limesalts exceeding the operational limit of 0.100g CaO/100ds and thick often below 8.0 pH during such events. These under-alkalization events were related to sudden changes in beet quality and soda ash addition adjustment lag relative to juice quality. Similar adjustment lag was also noted when beet quality again improved and soda ash was not simultaneously reduced relative to the juice in process resulting in over-alkalization of the juice. Over-alkalization is noted as lower limesalts concentration in hard thin juice along with higher thick juice pH and color rise across evaporation. Such events may be significantly reduced or even eliminated by pH control of the addition of alkalizing agent as noted in the "double alkalization" control described above.

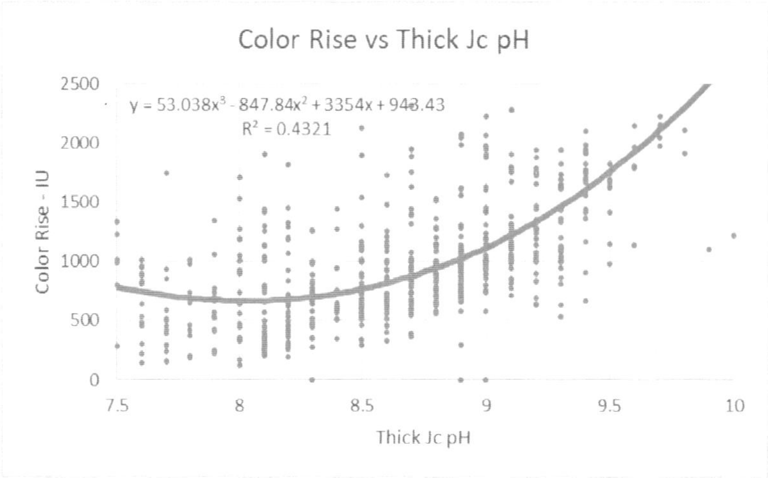

Figure 2

Figure 3 shows the difference in juice pH stability between a "flow based / manual ratio" control approach to soda ash addition and the "automatic pH control" of soda ash addition referred as double alkalization. During the period shown in Figure 3, hard juice limesalts control averaged 0.027g CaO/100ds for the entire period while the standard deviation in 2nd carbonation limesalts increased three-fold from the "manual control" period to the "pH auto control' period with virtually no change in soft juice limesalts. It is quite easy to see the improvement in the pH control stability in spite of increased variation in process juice quality.

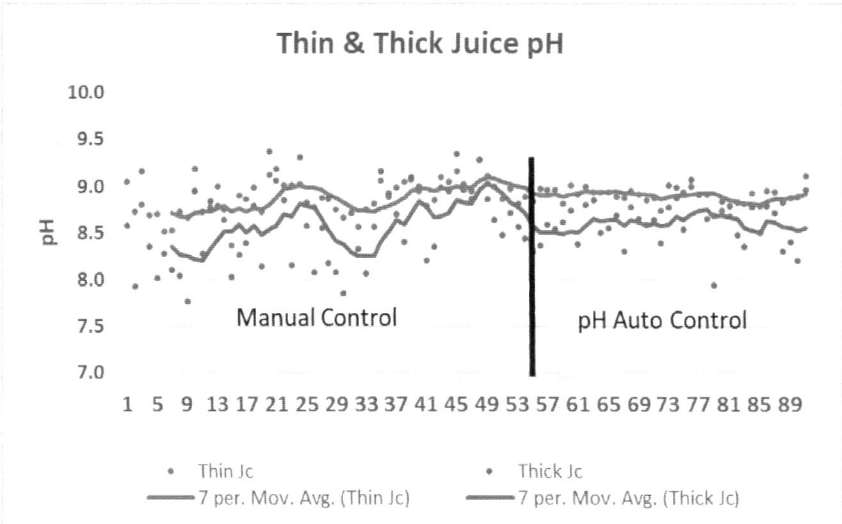

Figure 3

Figure 4 shows the 2nd carbonation, hard juice and soft juice limesalts concentrations for the same period shown in Figure 3. With pH control of soda ash addition, even with a much larger variation in 2nd carbonation limesalts concentration, the hard juice limesalts concentration did not vary significantly. This is due to 2nd carbonation being operated at optimum alkalinity and the soda ash addition to the juice in process automatically adjusted relative to the variation in the pH of the 2nd carbonation effluent. As a result, process adjustment lag is eliminated improving the overall control of the alkalization process.

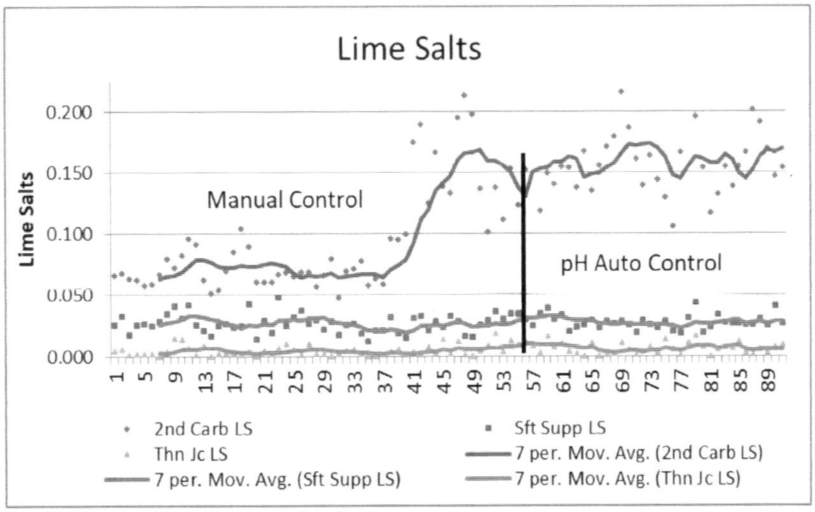

Figure 4

In short, optimal juice alkalization usually always results in sufficiently low limesalts concentration in hard juice for supply to the WAC softening cells and also in thick juice having ideal pH in roughly the range of 8.5+/- along with minimum color rise across juice concentration. The pH profile across the sugar end is also relatively stable exhibiting minimal pH drop from standard liquor to molasses relative to the beet in process.

WAC Softening Performance and Capacity Observations

Assuming that WAC service cycles and resin regeneration parameters are set within proscribed operating guidelines, certain operational performance characteristics may be observed. These characteristics are influenced by the operational environment to which the WAC system is subjected. However, observations of a number of installations in different operating environments have indicated that certain characteristics appear to manifest regardless of the operating environment.

The first of these observations is related to the expected reduction in limesalts across the WAC cells. In general, the reduction is a function of the hard juice supply limesalts concentration. In theory, since a soft juice limesalts concentration of zero is the minimum, the reduction across the WAC cells is directly proportional to the hard juice supply limesalts concentration. Thus, a lower hard juice supply limesalts results in less reduction across the WAC cells and, potentially, a somewhat lower soft juice limesalts concentration. The actual reduction is influenced by a number of factors including:

- The degree to which hard juice limesalts is controlled within operating guidelines.
- The amount of Ca^{++} "overloading" to which the resin is subjected.
- The degree to which the resin bed is subjected to fouling of the void volume causing channeling.
- The actual "working" capacity of the resin.

Each of these factors have an individual effect and are generally additive in their influence on WAC performance. All of these factors, with the possible exception of resin capacity which is influenced significantly by the other three factors, are controllable and significant with respect to system performance.

Figure 5 shows an example of the influence of hard juice limesalts on soft juice produced from the WAC cells. Virtually every change in hard juice limesalts concentration to the WAC cells, over the 72 operating days noted in the trend plot, resulted in a corresponding change in soft thin juice limesalts.

This is especially the case whenever the hard juice supply to the WAC units approached or exceeded the operational limesalts limit of 0.100g CaO/100ds.

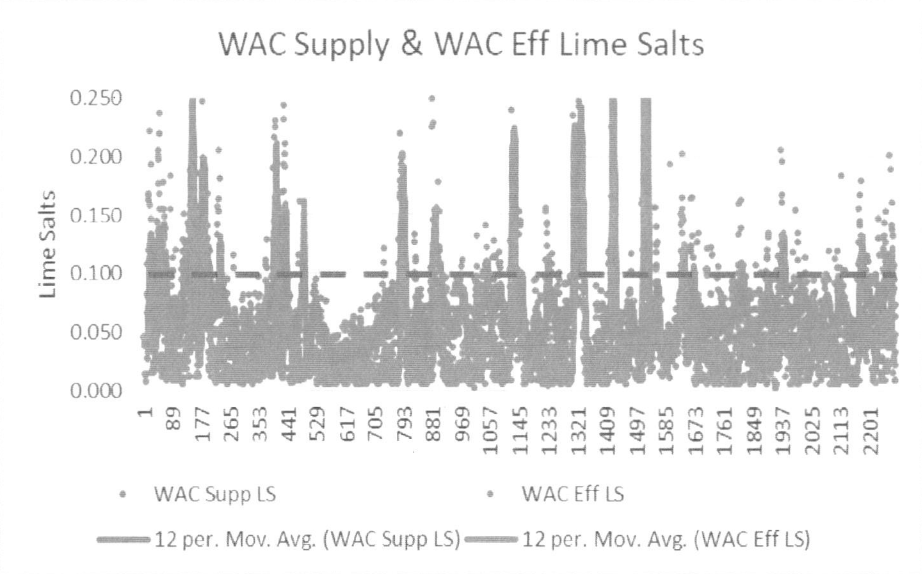

Figure 5

The average degree of reduction in limesalts across the WAC cells can be determined by comparing the limesalts reduction to the hard juice limesalts supplied to the WAC cells. The relationship will vary from one installation to another depending on the degree to which the hard juice supplied is consistently within acceptable operational limits. Figure 6 shows this relationship for the same installation as that in Figure 5. The scatter in the data points is largely the result of resin overloading due to frequent and/or sustained excursions in hard juice supply limesalts above 0.100g CaO/100ds.

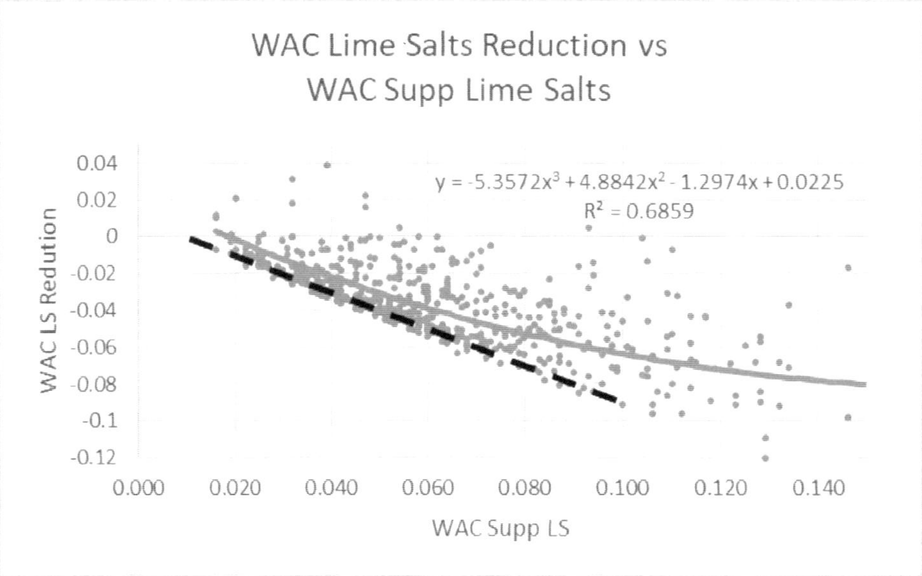

Figure 6

This relationship is quite important in setting a suitable target hard juice limesalts concentration to achieve the desired concentration of limesalts in soft thin juice. For example, from the illustration in Figure 6, WAC juice supply limesalts at 0.060g CaO/100ds gives an average reduction (blue trend line) of approximately 0.038 limesalts across the WAC units yielding soft juice having 0.022 limesalts: the approximate targeted soft juice limesalts concentration for this installation.

It is worth noting the "theoretical" operation that is shown by the minimum data points (black dashed line) on the scatter diagram in Figure 6. This is the best possible performance that may be expected from the system operating with little deviation from optimal performance. If such optimal performance could be routinely achieved, then any hard juice limesalts concentration below 0.100g CaO/100ds would produce a soft thin juice having approximately 0.010g CaO/100ds. Of course, it is not likely that such optimal performance always occurs in actual practice. It is also important to note that the concentration of data points on the "optimal trend" increases as the concentration of hard juice limesalts decreases. The degree of "scatter" in the operation depends on the degree of "scatter" in the WAC supply hardness along with the degree to which hard juice supply limesalts exceeds the operational limit of 0.100g CaO/100ds.

Whenever the WAC supply limesalts concentration exceeds the operational limit, the capacity of the resin is at least temporarily affected. Figure 7 shows what generally occurs during and after a typical excursion in WAC supply limesalts.

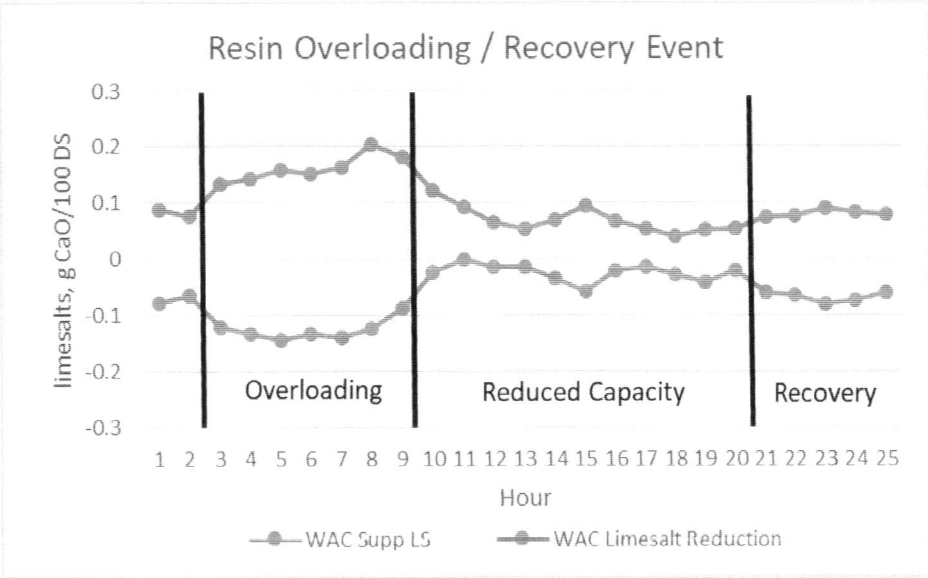

Figure 7

During the "Overloading" phase, the resin removes an increased amount of Ca^{++} (reduction across WAC cells exceeding 0.100g CaO/100ds) until it is effectively fully loaded. At that point the resin is essentially unable to remove any more Ca^{++} from the juice and enters a "Reduced Capacity" phase. During this phase the reduction in limesalts across the WAC cells is reduced to near zero (may actually give up Ca^{++} to the WAC effluent juice). This condition is largely responsible for the significant "scatter" in the data points shown in Figure 6. When the hard juice limesalts is once again managed below the operational limit, the resin then enters a "recovery" phase and eventually regains its operating capacity over the course of multiple regeneration cycles. In this example, the WAC system required approximately 12 hours to return to acceptable operational capacity after the excursion above the operational limit of 0.100g CaO/100ds hard juice limesalts. This is a bit longer than the length of time required to complete one regeneration of all WAC cells in this WAC installation.

To achieve design performance of the WAC softening system, overloading of the resin due to hard juice supply limesalts exceeding the operational limit must be avoided. Such operational control integrity is necessary to insure that the resin capacity can achieve stable and maximum limesalts reduction across the WAC cells.

Resin capacity analysis more or less confirms the influence of frequent resin overloading. Table 1 shows the resin capacity analysis for several WAC systems currently in operation. Factory E sampled the resin just after a series of relatively significant overloading events.

	Resin Capacity				
	New	Half Camp	Full Camp	Fctry Regen	Design
Factory A	4.3	4.1	3.9		1.5
Factory B	4.3	3.8	3.7		1.5
Factory C	4.3	3.7	3.6		1.5
Factory D	4.3	3.7	3.2		1.5
Factory E	4.3	3.2 (30 days)		2.6 (30 days)	1.5

Table 1

The analysis of the Factory E "Fctry Regen" sample is only 40% of the capacity of new resin and also only about 81% of the capacity of the HCl pretreated portion of the same resin sample. The "Fctry Regen" sample was not subjected to the usual "acid stripping" before capacity measurement. Even so, the acid stripped resin sample is also lower than the Factory A-D samples which were taken after approximately 2 months of operation whereas the Factory E sample was taken after only one month of operation during which Factory E frequently operated with hard juice supply to the WAC cells exceeding the operational limit of 0.100g CaO/100ds. The comparatively lower capacity for the Factory E resin may suggest increased fouling of the resin due to the operating conditions to which it was subjected.

Operating Targets for Juice Alkalization

The primary goal of juice alkalization, with respect to the operation of WAC juice softening, is the control of the limesalts concentration in the hard juice supply to the WAC cells. As noted, the normal variation in 2nd carbonation effluent limesalts and the resulting variation in hard juice limesalts related to the control of alkalizing agent dosage must be taken into consideration in setting the average hard juice limesalts target. Depending on the normal variation of processing variables, the targeted average limesalts concentration must be adjusted so there are NO excursions in limesaltss above 0.100g CaO/100ds. Such operating performance may be difficult to attain if there are large fluctuations in 2nd carbonation limesalts concentration in the absence of fully automatic control of soda ash addition. If limesalts excursions exceeding the design concentration limit occur, the capacity of the WAC resin will be affected as well as the reliability of the limesalts concentration of the WAC effluent soft juice.

The degree to which the process is directly managed with online process measurement has a critical influence on the stability of the process. One possible method has been described herein and has shown positive results in its limited deployment where large variation in beet quality and 2nd carbonation effluent limesalts were experienced. One extremely important aspect of the performance of the WAC softening system is to minimize the variation in hard juice limesalts concentration and to avoid any excursion in limesalts concentration above the operational limit of hard juice supply to the WAC cells.

Many system operators seem to arrive at a hard juice limesalts target that is significantly below the operational limit. The hard juice limesalts target often depends on the variation in hard juice limesalts experienced in routine operation and the soft juice limesalts targeted by that operation. Table 2 gives approximate general guidelines followed by many operations in the USA.

Limesalts Targets: g CaO/100ds	
Hard Jc	Soft Jc
0.030 - 0.040	0.000 - 0.005
0.045 - 0.055	0.008 - 0.015
0.060 - 0.070	0.020 - 0.025

Table 2

A secondary, but no less important consideration in establishing juice alkalization targets are the pH and color of thick juice and pH stability in the sugar end. Ideally, as shown in Figure 2, a thick juice pH should be targeted relative to a minimum color rise in juice concentration. The actual color rise noted at optimal thick juice pH will be influenced by the type of evaporator employed and the quality of the beets in process. Under most conditions a final thick juice pH in the range of 8.5+/-0.03 accompanied by a slight pH drop across the evaporator in the range of 0.02-0.05 pH is usually associated with the lowest color rise.

If the juice is alkalized with soda ash to yield the optimum thick juice pH and there is no additional juice alkalization (caustic addition) downstream of the WAC softening cell, the average limesalts in hard thin juice will likely be in the range of 0.035 to 0.060g CaO/100ds and fully suitable for supply to the WAC cells. WAC effluent soft juice will commonly be in the range of 0.010+/- 0.002g CaO/100ds or less if resin overloading is avoided. In cases where a higher soft juice limesalts concentration is desired, a small amount of hard thin juice may be bypassed to achieve the desired limesalts concentration in soft thin juice.

Unless the juice is under-alkalized with soda ash prior to entering the WAC cells, it should not be necessary to add additional alkali after the WAC units to achieve ideal conditions for juice concentration and sugar end pH control. However, under certain conditions where there is very low natural alkalinity in beets processed or where a higher limesalts concentration in hard thin juice is desired, there may be a post WAC alkali adjustment requirement. This may especially be the case if the control of soda ash addition is less than ideal or not automatically controlled as described earlier and the juice is being deliberately under alkalized in the 2nd carbonation operation.

In Summary

To summarize the influence of thin juice alkalization on WAC thin juice softening performance and juice quality, the following points are emphasized:

1. Juice alkalization must be managed to insure that the hard juice supply to WAC cells does not exceed 0.100g CaO/100ds. Both over and under alkalization of juice is to be avoided.
2. A 2-step pH based control of the alkalization process provides stable on-line control of hard juice limesalts and thick juice pH. Such control is especially important where large excursions in juice quality (2nd Carb limesalts) are experienced.
3. If hard juice exceeds the operational limit of 0.100g CaO/100ds, resin capacity is temporarily reduced resulting in increased soft juice limesalts and reduced resin capacity. Resin capacity is generally not restored until regeneration of all cells in the system is completed.
4. Juice alkalization management targeting optimal thick juice pH and color behavior during juice concentration generally results in limesalts concentration within acceptable limits in the hard juice supplied to WAC softening cells.

BOLTON, MELVIN D.[1,2*], SPANNER, REBECCA[2] AND SECOR, GARY A.[2], [1]USDA-ARS, Red River Valley Agricultural Research Center, Fargo, ND, USA; [2]Department of Plant Pathology, North Dakota State University, Fargo, ND, USA. **Use of genome-wide association to identify mutations involved with DMI-resistance in *Cercospora beticola*.**

Management of Cercospora leaf spot relies on the application of fungicides. A widely-used class of fungicides is the sterol demethylation-inhibiting (DMI) fungicides that inhibit the cytochrome P450 14α-demethylase CYP51. CYP51 is an enzyme that catalyzes a key step in fungal ergosterol biosynthesis, which is required for fungal survival. Quantitative resistance to DMIs has emerged in *C. beticola* populations due to their widespread use. To identify mutations responsible for DMI resistance, a genome-wide association study was undertaken using 200 isolates harvested from sugarbeet fields in the Red River Valley of ND and MN. Approximately half of the isolates were DMI-sensitive (EC_{50} value <1) while the other half were DMI-resistant (EC_{50} value $\geq$1). After Illumina whole genome resequencing, genome wide association identified three loci highly correlated with DMI-resistance. The results of this study will be reported.

WEILAND, JOHN J.[1*], NEUBAUER, JONATHAN[1] AND BOLTON, MELVIN D. [1]USDA-ARS, Red River Valley Agricultural Research Center, Fargo, ND, USA. **Commercial sugar beet in Minnesota possessing the Rz2 gene and testing positive for BNYVV.**

Sugar beet production in Minnesota was under threat from the emergence of Rhizomania disease in the 1990s and the inclusion of Rz1 gene for resistance to the disease in subsequent approved varieties proved efficacious in producer fields. Defeat of Rz1 in many regions by resistance-breaking (RB) isolates of BNYVV, causal agent of rhizomania disease, prompted addition of a second resistance gene, Rz2, to the defense arsenal in an effort to retain economic viability to sugar beet production where needed. As a proactive measure, the majority of producers in Minnesota currently grow commercial hybrids possessing both genes; additional genes may be present as part of the resistance selection process implemented by the seed industry. In an effort to establish a baseline for the detection of BNYVV in field-grown sugar beet possessing the combined Rz1 +Rz2 resistance package, beet roots of such varieties as well as several possessing Rz1 alone were obtained from production fields in the American Crystal, Min-Dak Farmers, and the Southern Minnesota Beet Sugar Cooperative regions of Minnesota. Beet root morphology ranged from those exhibiting classic Rhizomania disease to healthy looking roots with several of the latter nevertheless exhibiting occasional hairy root proliferation in the root furrow. Testing of the lateral rootlets and root hairs by ELISA yielded positive confirmation of the presence of BNYVV in the tissue for most roots tested while several healthy roots tested negative for the virus. The molecular typing of the BNYVV obtained from test-positive roots with respect to genomic regions implicated in RB ability will be presented. Factors combining with the discovery in the impact on future yields will be discussed.

WEILAND, JOHN J.[1*], SHARMAPOUDEL, ROSHAN[2], SECOR, GARY A.[2] AND BOLTON, MELVIN D. [1]USDA-ARS, Red River Valley Agricultural Research Center, Fargo, ND, USA. [2]Department of Plant Pathology, North Dakota State University, Fargo, ND, USA. **A novel satellite virus discovered during a bait assay for soilborne viruses of sugar beet.**

Next generation sequencing has emerged as a powerful diagnostic tool to assist the plant pathologist in the detection and diagnosis of new and known disease agents. In an effort to better characterize the milieu of soilborne viruses present in conjunction with BNYVV in association with rhizomania disease, sugar beet seed of a rhizomania-susceptible variety was sown into soil typed for inducing severe rhizomania disease. At 5 weeks post planting, seedlings were harvested and the roots carefully washed prior to an extractive treatment which enriched for the presence of viruses within the sample. Electrophoresis of the RNA extracted from the enriched sample revealed an unusually abundant RNA species (~1.2 kilobases in size) in addition to standard ribosomal RNA contaminants of the preparation. Cloning and subsequent sequencing of the RNA species indicated it to be a previously-unknown satellite virus weakly related to satellite maize white line mosaic virus and other satellite viruses associated with ‘helper’ Necroviruses. Future objectives will be to identify the helper virus associated with the novel satellite virus among the many viruses detected within the analysis and to produce infectious cDNA clones of both helper and satellite for pathogenicity studies. Infectivity studies of this nature ultimately will lead to an understanding of any contribution of this virus/satellite combination to rhizomania or other sugar beet diseases.

Sugar Beet Breeding Progress Update of Cercospora Tolerant Varieties in North America
ASSBT Conference, February 27, 2019

JAY MILLER Director of Product Management

PATRICK O'BOYLE Plant Breeder

MARGARET REKOSKE Director of Breeding and Product Development

BETASEED INCORPORATED

BETASEED INCORPORATED

Cercospora Leaf Spot

- Cercospora leaf spot (*Cercospora beticola*) is a fungal disease that occur in many sugar beet producing areas of the United States.
- Infection often results in significant reductions in yield and sucrose such as in Southern Minnesota in 1981 where yield and sugars were reduced up to 50%.
- The principal means of controlling losses due to infection is through tolerant varieties and/or fungicides.
- In the last 50 years the pathogen has shown the ability to overcome fungicides.

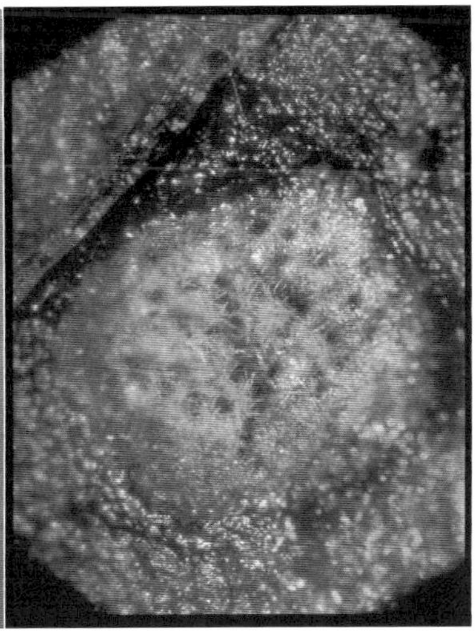

BETASEED INCORPORATED

Cercospora Leaf Spot Incidence

- Affects about 3/4ths of the North American sugar beet growing area
  - MN, ND, MI are the major regions
- Several severe infestations in North America have occurred in the last 40 years – early 1980's, late 1990's and most recently in 2016

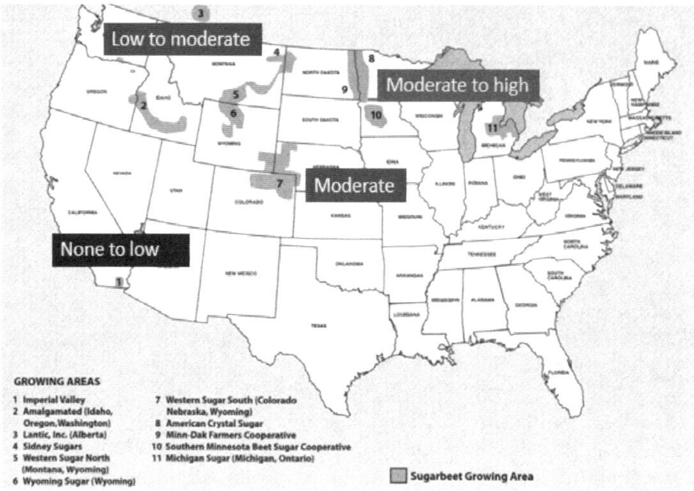

BETASEED INCORPORATED

Management Of Cercospora Leaf Spot

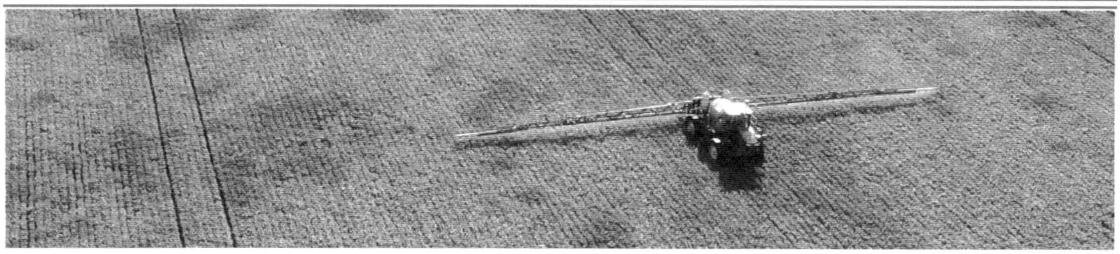

Two main methods of controlling Cercospora

- Application of fungicides – monitoring systems are critical for timing the application
- Tolerant varieties

Other contributing factors

- Variety approval requirements
- Agronomic practices such as crop rotation, weed control, tillage methods, etc.

Variety Approval Requirement for Cercospora Changes Across N.A. Markets

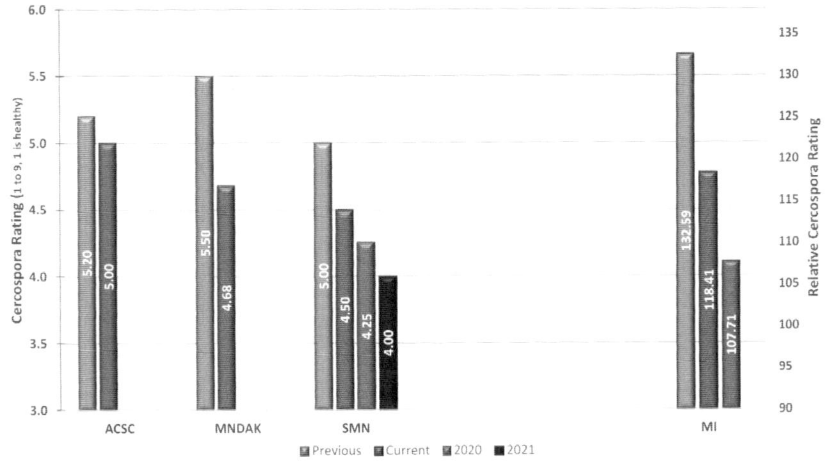

Historic U.S. Beet Sugar Production (Lbs/A)

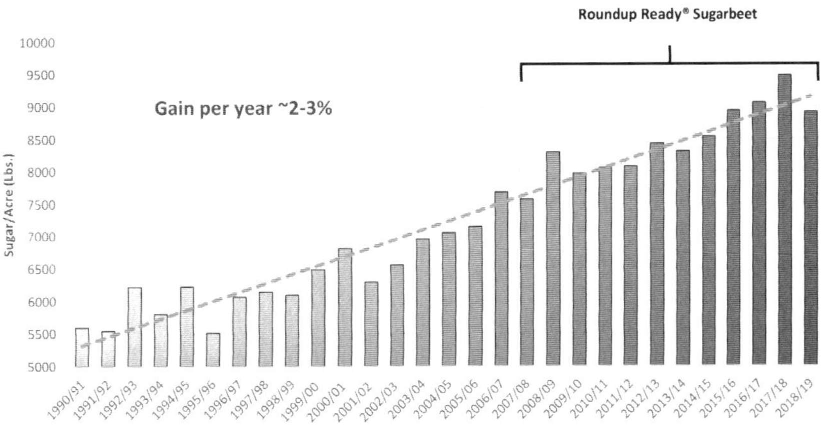

Source: USDA, National Agricultural Statistics Service, Crop Production; Farm Service Agency, Sweetener Market Data; World Agricultural Outlook Board, WASDE; Economic Research Service, Sugar and Sweetener Outlook.
https://ers.usda.gov./data-products/sugar-and-sweeteners-yearbood-tables/

Breeding For Cercospora Tolerance

- In the 80's and 90's the best Cercospora tolerant varieties were at least 10% lower in performance - tons and/or sugar content
- Strong genetic variation for tolerance exists
- Breeding challenge - inverse correlation between Cercospora tolerance and performance (sugar yield per unit of land)
- Breeding strategy and methodology has changed over the last 25 years

General Breeding Tendency Of Cercospora Tolerance Versus Performance

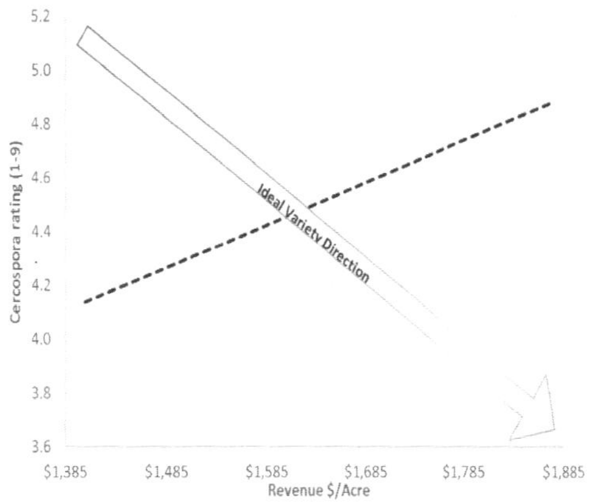

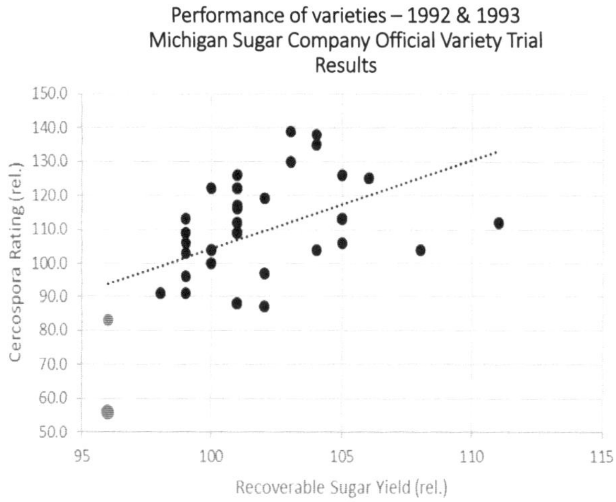

BETASEED INCORPORATED

SMBSC Official Trial Data 2011-2018*

Betaseed Inc. Genetics

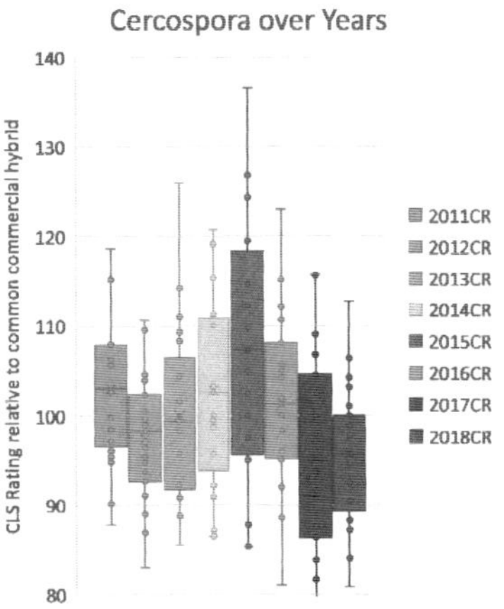

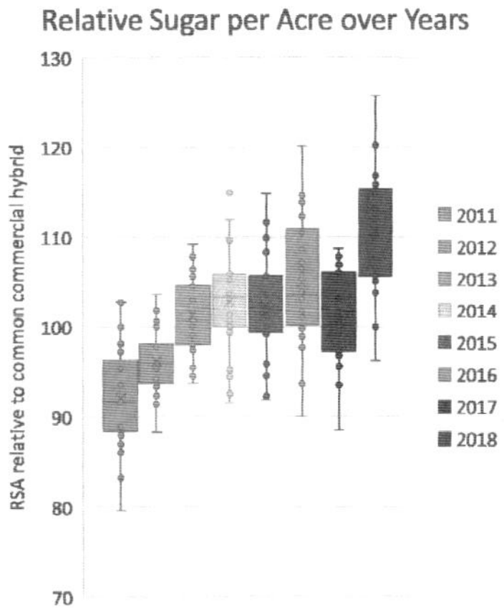

*Data used with permission from Southern Minnesota Beet Sugar Cooperative

BETASEED INCORPORATED

Last 6 Years Minn-Dak OVT Results*

Cercospora Score versus Financial Income

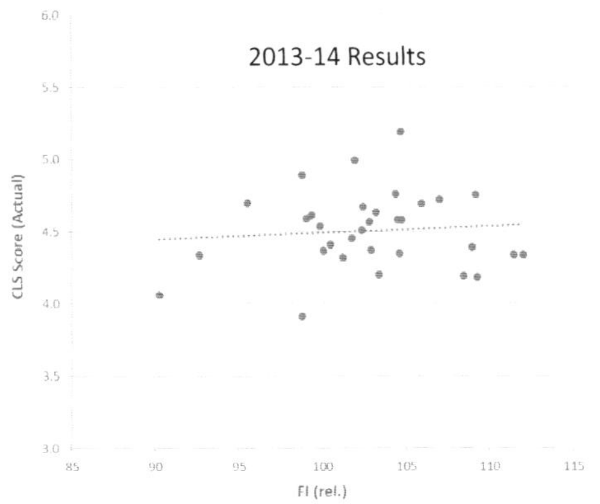

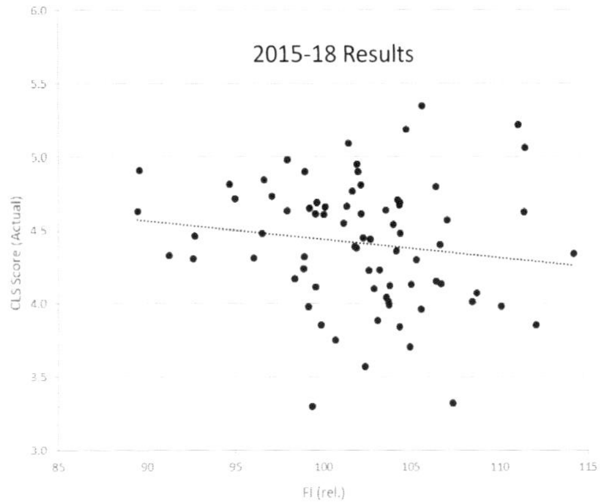

*Data used with permission from Minn-Dak Farmers Cooperative

BETASEED INCORPORATED

Milestones In The Development Of Cercospora Tolerant Hybrids

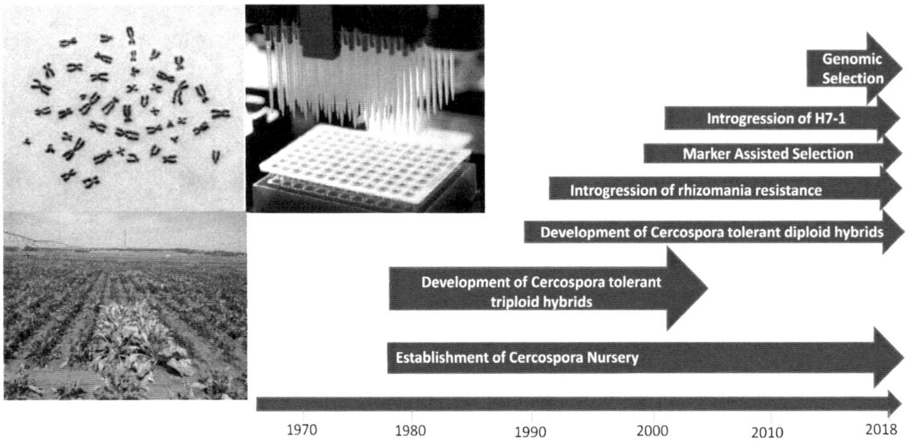

BETASEED INCORPORATED

Advanced SCA Hybrids 2010-2018

Betaseed Inc. Genetics - Red River Valley

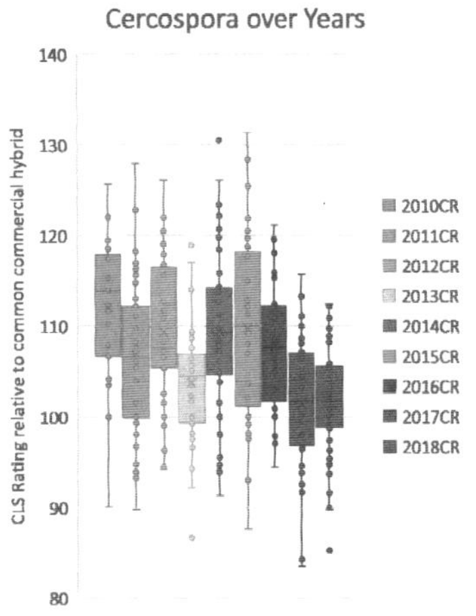

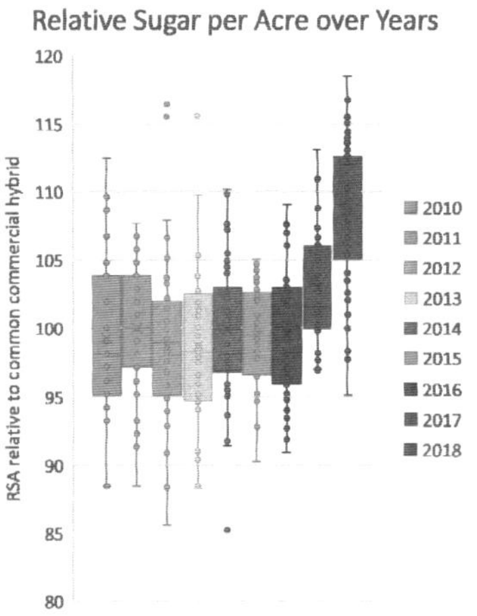

BETASEED INCORPORATED

Performance Of Cercospora Tolerant Material

Michigan Proprietary Yield Trial (2017 and 2018)

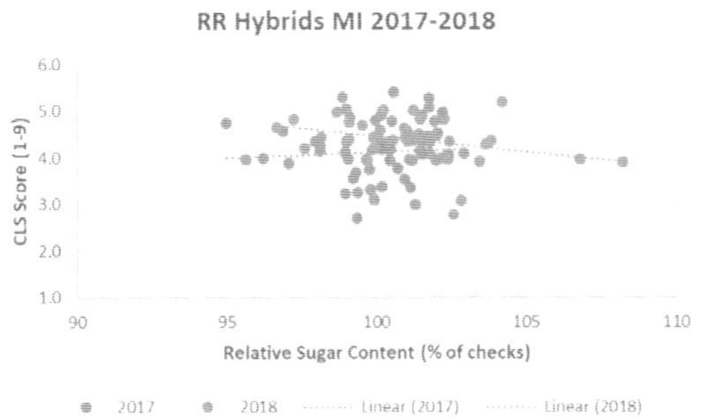

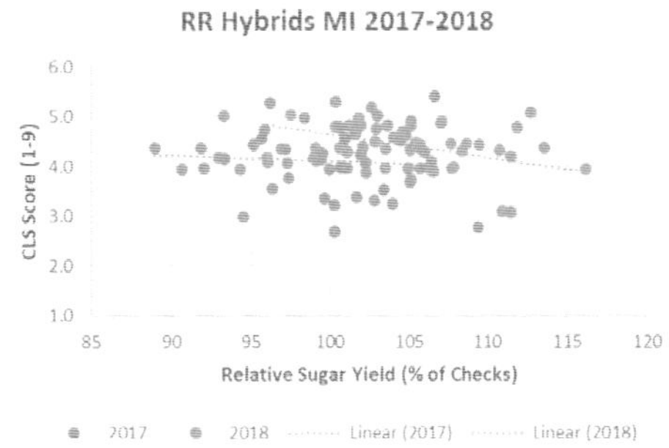

- Opportunity exists to select highly tolerant hybrids that are comparable in performance to less tolerant hybrids

BETASEED INCORPORATED

Development Of Cercospora Tolerant Hybrids

- Evaluation of thousands of new potential lines annually in our Cercospora nurseries.
- Development of molecular tools to facilitate quicker breeding gain – GS, QTL identification.
- Addition of important traits into highly Cercospora tolerant hybrids: nematode, rhizomania, root rots, root aphid, herbicide tolerance.

BETASEED INCORPORATED

Summary

- Breeding advances have provided better Cercospora tolerant varieties
- The inverse correlation of Cercospora tolerant to sugar yield still exists but to a lesser extent than 25 years ago
- Some highly Cercospora tolerant varieties show comparable performance and higher revenue to non-tolerant hybrids even under no infection
- Important Cercospora tolerant materials are widely used in North America
- Over the last 10 years the gain in sugar yield is about 2% to 3% per year including Cercospora tolerant hybrids

BETASEED INCORPORATED

Future

BETASEED
INCORPORATED

Acknowledgement And References

- Acknowledgements
  - Data used with permission from Research Groups of Michigan Sugar Company, Minn-Dak Farmers Cooperative and SMBSC

- References

  - Compendium of Beet Diseases and Pests Second Edition (2009). The American Phytopathological Society.
  - Miller, J., Rekoske, M., and Quinn, A. (1994). Genetic resistance, fungicide protection and variety approval policies for controlling yield losses from Cercospora leaf spot infections. J. Sugar Beet Res. 31, 7–12.
  - Smith, G. A., and Gaskill, J. O. (1970). Inheritance of resistance to Cercospora leafspot in sugarbeet. J. Am. Soc. Sugar Beet Technol. 16, 172–180.
  - Vogel J, Kenter C, Holst C and Märländer B (2018) New Generation of Resistant Sugar Beet Varieties for Advanced Integrated Management of Cercospora Leaf Spot in Central Europe. Front. Plant Sci. 9:222.

BRANTNER, JASON R.* and ASHOK K. CHANDA, Department of Plant Pathology, University of Minnesota, Northwest Research and Outreach Center, Crookston, MN 56716. **Accurate diagnosis of sugarbeet root diseases is critical for driving disease management decisions.**

ABSTRACT

Accurate diagnosis of soilborne sugarbeet root diseases is not only critical for implementing effective disease management strategies but also useful in documenting prevailing diseases in a given year. Treatment applications in the current growing season and selection of variety, seed treatment, and fungicide applications for the next time the field is planted to sugarbeet all depend on knowing what pathogen(s) are present in the field. The objective of our field sample diagnosis is to identify root pathogens and provide information to growers.

In the last ten years (2009-2018) the sugarbeet pathology lab at the University of Minnesota, Northwest Research and Outreach Center has diagnosed samples from 1088 fields from across the sugarbeet growing areas of Minnesota and North Dakota. Diagnosis is performed by gathering and assessing pertinent information from sample submitters followed by isolation of pathogens from infected tissues where appropriate and microscopic observation.

The most common diseases were damping-off and/or adult root rots caused by *Rhizoctonia solani* (593 samples = 55%) and *Aphanomyces cochlioides* (332 samples = 31%). Diseases caused by *Fusarium oxysporum* or *F. secorum* have been found in 56 samples (5%) during this period. Other pathogens appearing infrequently are *Pythium spp*., and *Verticillium dahliae*. Chemical issues, mostly herbicide carryover or drift, have been suggested in 31 samples (3%). In the last four years, both *R. solani* and *A. cochlioides* have been isolated from 40 of 355 samples (11%) corroborating an increasingly common observation of co-existence of both diseases in many fields. During the 10-year period, samples with Rhizoctonia diseases have been consistently the most common, while those with Aphanomyces have fluctuated from year to year (Fig. 1), depending on the soil moisture. Samples with Fusarium diseases peaked in 2013, but have remained low since then (Fig. 1) due to grower awareness and selection of tolerant varieties.

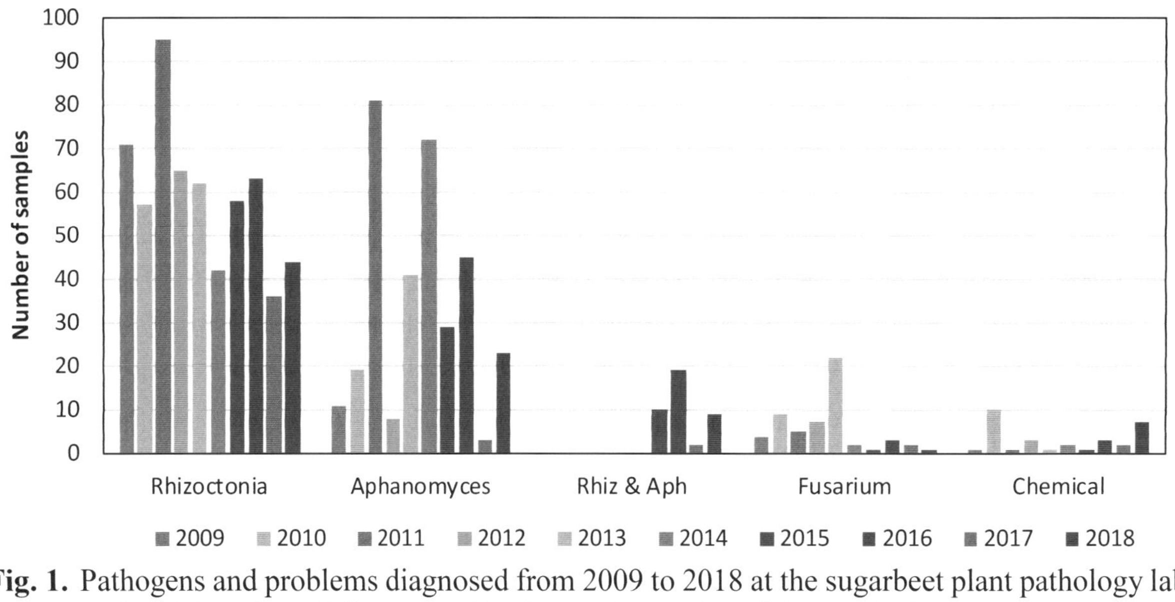

Fig. 1. Pathogens and problems diagnosed from 2009 to 2018 at the sugarbeet plant pathology lab at the University of Minnesota, Northwest Research and Outreach Center.

There is much anecdotal evidence of misidentification of root diseases leading to inappropriate variety and treatment selection. Proper identification of root diseases in the field involves more than simple above-ground observation and should include digging up several intact roots with varying levels of symptoms around an infection focus or problem area, removing soil to better observe symptoms, and for some pathogens, cutting the roots to observe vascular tissue for presence of discoloration. At times, symptoms caused by *R. solani* and *A. cochlioides* can be difficult to discern and isolation of the pathogen or its DNA by a qualified lab is necessary. Proper samples for this procedure should include 4-8 roots with clear symptoms of active infection alongside healthy tissue. Completely rotten roots are usually not helpful as secondary organisms have taken over by that time. Accurate identification of pathogens in sugarbeet fields through proper sampling and diagnosis will give growers valuable information to make better disease management decisions including taking measures to prevent spread of soilborne pathogens from infested fields to clean fields.

Application of Near-Infrared Spectroscopy (NIRS) for the Detection of Sucrose and Betaine in Streams Surrounding the Molasses Desugaring by Chromatographic Process (MDC)

Carol L. Wambolt*, Doris C. Fleming, Alla A. Bagramyan and Larson H. Caldwell
Amalgamated Research, LLC.,
2531 Orchard Drive East, Twin Falls, Idaho 83301

Since the mid-1980s, industrial chromatographic separators have been utilized in U.S. beet sugar factories for molasses desugarization. This technology has proven to be an advantageous extension of the factory to increase sugar yield as well as aid in the recovery of other valuable components such as betaine. A simplified layout of a typical coupled loop separator is shown below in Figure 1. It consists of two units, called loops. Softened and filtered molasses, along with water, are fed into loop1. Two fractions are then produced; the material moving ahead in the profile or the upgrade fraction (sucrose, salts, color and some organics) and the slower moving compounds or the betaine fraction (betaine and amino acids). The upgrade fraction from loop 1 serves as the feed to loop 2. In loop 2, the salts, ionized organic compounds and color (raffinate fraction) are separated from the sucrose (extract fraction). The separator is a very dynamic and complex system and to properly "tune" or optimize its operation, rapid analytical methodology is needed to monitor sucrose, betaine and refractometric dissolved solids (RDS) levels around the various separator streams.

Figure 1. Coupled Loop Molasses Separator

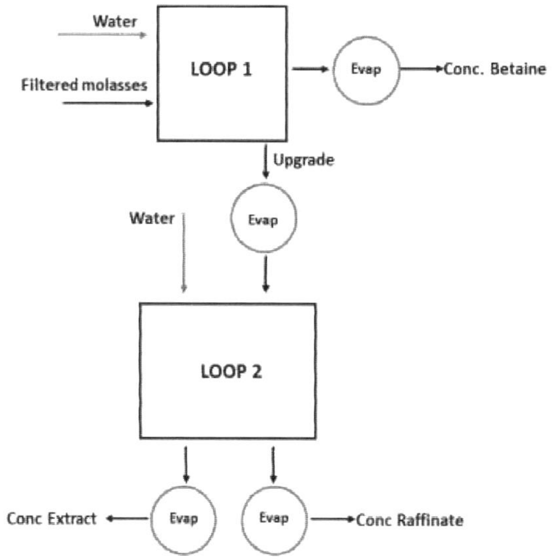

The analytical group at Amalgamated Research, LLC. has developed methodology, based on Near-infrared spectroscopy (NIRS), to accurately predict sucrose, betaine and RDS values across a wide range of concentrations, allowing for effective monitoring of the molasses desugaring separator. NIR spectroscopic techniques offer several advantages to traditional analytical methods. The NIR technique is a rapid method, producing data in 30-60 seconds, rather than an hour, requires zero sample preparation and uses no chemicals or consumables. The predictive performance of this technique will be discussed.

Introduction:

In 1800, Fredrick William Herschel was credited for discovering the region of the near-infrared spectrum, in his work "Experiments on the Refrangibility of the Invisible Rays of the Sun." This discovery was largely ignored for 150 years and was referred to as a "sleeper among spectroscopic techniques." The first reported application, utilizing NIR technology, was in the 1950s, mainly for measurement of moisture in hazardous materials. However, it wasn't until the 1970s that NIR techniques started to rapidly progress. The application of NIR to the food and agricultural industry began with Karl Norris, an engineer at the US Department of Agriculture, who truly recognized the potential of this technique. Norris was able to expand NIR technology into industrial practice by developing instrumentation with improved optics and electronics, coupled with computer algorithms capable of processing complex spectroscopic data. From the 1970s onward, NIRS has gained wide acceptance in many and varied industries. Typical applications include use in the pharmaceutical industry for raw material testing, product quality control and process monitoring, food (dairy, meat, grain, etc.), agrochemical quality control, petrochemical, cosmetics, medical diagnostics (ie. blood sugar and pulse oximetry), etc.

The American Society of Testing and Materials (ASTM) defines the NIR region of the electromagnetic spectrum as the wavelength range from 780 to 2,526 nm. When light energy interacts with a material, energy is absorbed at resonant frequencies associated with the atomic and molecular interactions of the material. The most prominent absorption bands occurring in the NIR region are related to overtones and combinations of fundamental vibrations of -CH, -NH, -OH and -SH functional groups.

There are three parts of an NIR spectral region. When incident NIR radiation strikes a sample, the following three absorbance phenomena take place: Transflectance, Transmission or Reflectance.

Transflectance: The 720 to 1100 nm region is most suited to transflectance through thick samples such as seeds, slurries, liquids and pastes. The absorption bands are due to 3rd overtones of the fundamental stretch bonds in the Mid-infrared (MIR) region.

Transmission: The 1200 to 1850 nm region is used for transmission through liquids and films and samples with high water content. The absorption bands are due to the 1st and 2nd overtones of the fundamental stretch bonds in the MIR region.

Reflectance: The 1850 to 2500 nm region is used for making diffuse reflectance measurements of ground or solid materials. The absorption bands are due to combination bands, such as the -CH stretch and bend combination bands.

Near-infrared spectroscopy (NIRS) is a rapid, nondestructive analytical technique that provides chemical and physical information on virtually any sample matrix. NIRS in combination with chemometrics, opens many options for development of non-invasive, rapid measurement techniques used for both qualitative and quantitative purposes.

Materials and Methods:

Sucrose, Betaine and RDS samples

The samples for this study were collected at The Amalgamated Sugar Company, LLC. Twin Falls and Nampa factories. The eleven separator streams had average sucrose, betaine and RDS concentrations ranging from 0.39 to 60.2, %m/m, 0.4 to 50.0, %m/m and 4.29 to 81.34 %m/m, respectively. Each model contained approximately 130-400 individual data points. Samples were kept refrigerated until analysis commenced. Aliquots of each sample were analyzed via the primary reference methods, gas chromatography (GC), high performance liquid chromatography (HPLC) or refractometry. The remaining sample portions were submitted for NIR scanning. All spectroscopic scans were performed at room temperature (~23 °C). Table 1 lists the separator sample streams selected for this study.

Table 1.

Model Designation	Separator Stream	Factory Location
T1	Virgin Molasses	Twin Falls
T2	Separator Feed Molasses	Twin Falls
T3	Concentrated Betaine	Twin Falls
T4	Dilute Upgrade Molasses	Twin Falls
T5	Dilute Betaine	Twin Falls
T6	Amino Acid Cut	Twin Falls
T7	Concentrated Upgrade Molasses	Twin Falls
T8	Concentrated Extract	Twin Falls
T9	Concentrated Raffinate	Twin Falls
T10	Trains A and B_Dilute Extract	Twin Falls
T11	Trains A and B_Dilute Raffinate	Twin Falls
N1	Virgin Molasses	Nampa
N2	Concentrated Betaine	Nampa
N3	Betaine Molasses	Nampa
N4	BUS_Concentrated Betaine Feed	Nampa
N5	BUS_Upgrade Betaine Dilute	Nampa
N6	BUS_Upgrade Betaine Concentrated	Nampa
N7	BUS_Reject	Nampa
N8	Loop 1 Trains A&B_Separator Feed Molasses	Nampa
N9	Loop 1 Trains A&B_Upgrade Molasses	Nampa
N10	Loop 1 Trains A&B_Betaine Cut	Nampa
N11	Loop 1 Trains A&B_Amino Acid Cut	Nampa

*BUS = Betaine Upgrade Separator

Instrumentation

An Agilent Gas Chromatograph (GC), Waters high performance liquid chromatography (HPLC) system, Rudolph Research Analytical J157 Refractometer and FOSS NIRS DS2500 NIR spectrometer were used to make the measurements for this study. The details of each instrument confuguration are listed below.

Gas Chromatography (GC)

Sucrose levels were quantified using the primary method, gas chromatography. An Agilent 6890N GC system, interfaced with a personal computer, was used to make these measurements.

GC system	
GC	Agilent 6890N (Operating software: OpenLAB CDS_EZChrom Edition)
Detector	Flame Ionization
Autosampler	7683 Series
Column	HP-5, 5% Phenyl 95% dimethylpolysiloxane, 30 m, 0.32mm, 0.25μm, 7 inch cage

Method parameters	
Sucrose std.	Sanding sugar (20 - 140 mg) Converted to its trimethylsilyl ether
Carrier gas	Helium_27 mL/min
Hydrogen to Det	25-30 mL/min
Air to Det	350-375 mL/min
Oven temp.	235℃
Inj/Det Temp.	250℃

High Performance Liquid Chromatography (HPLC)

Betaine levels were quantified using the primary method, high performance liquid chromatography. A Waters High Performance Liquid Chromatography (HPLC) system, interfaced with a personal computer, was used to make these measurements.

HPLC system	
HPLC	Waters (Operating software: Breeze 2)
Detector	Refractive index (RI 2414)
Pump	ISOpump 1515
Autosampler	Autosampler 2707
Column	Aminex HPX-87N (Na+ form), 300 X 7.8 mm, 9 μm particle size, 8% cross linkage, pH range 5-9

Method parameters	
Betaine std.	Betaine BioUltra (≥99.0% (NT))_Supplier: Sigma-Aldrich
Eluent	0.015 M Na_2SO_4
Flow rate	0.7 mL/minute
Column temp.	85℃
Detector Temp.	35℃

Refractometer

RDS levels were measured using the primary method, refractometry. A Rudolph Research Analytical J157 refractometer was used to make these measurements.

Refractometer:	
Refractometer	Rudolph J157
Measurement range	Brix 0 - 100
Accuracy	±0.01 Brix
Optical wavelength	589.3 nm (NaD line)

Near-infrared Spectrometer (NIRS)

NIR spectra were recorded with a FOSS NIRS DS2500 spectrometer, interfaced with a personal computer.

NIRS system:	
NIR Spectrometer	FOSS NIRS DS2500 (Operating software: ISIscan Nova)
Detector(s)	Silicon (400-1,100 nm) and Lead Sulfide (1,100-2,500 nm)
Sample cell	Slurry cup coupled with a gold reflector (Optical pathlength = 0.2 mm)

Method parameters:	
No. scans	32
Resolution (nm)	0.50
Spectrum range (nm)	400 - 2,500
Measurement mode	Transflectance
Final format	Log(1/R)
Optical bandwidth (nm)	8.75
Absorbance range	Up to 2 AU

Spectral data pre-treatments and regression method

The Near-Infrared (NIR) spectral pre-treatments and predictive model development were performed with WinISI (Version 4.7.0.14943). NIR spectra are affected by both physical (matrix inconsistencies) and chemical changes. Physical effects, such as sample particle size variation, can cause the baseline and maximum absorbance to vary. Spectral pre-treatments correct for these physical changes. Several pre-treatments were selected to minimize the standard error of prediction (SEP). The following spectral pre-treatments were applied: Standard Normal Variate (SNV), De-trending (DT), First Derivative and Smoothing.

Quantification models for sucrose, betaine and RDS were developed using Modified Partial Least Squares (MPLS) regression. MPLS is an improvement over the traditional partial least squares (PLS) regression. MPLS is an excellent regression technique for agricultural samples, that have a complex sample matrix, as it copes more effectively with non-analyte interference in multicomponent determinations and provides an estimate of the interference spectrum. The interference spectrum can be used to quantitatively correct the measured spectrum before determination of analyte concentrations; hence the accuracy of the NIRS prediction will be improved.

Results and Discussion;

The Near-Infrared (NIR) spectroscopic techniques allowed for the development of very good calibration models for the prediction of sucrose, betaine and RDS in eleven individual factory streams. Cross-validation (leave-one-out) was applied to all models. Cross-validation is a technique used in modeling to estimate the test error. A number of sample observations, referred to as validation sets, are partitioned from the training data set. After fitting the model on the training data set, its performance is measured against each validation set and averaged.

Two sets of samples were prepared for calibration and prediction. The calibration sets for each model consisted of 130-450 samples analyzed over a year. The validation sets consisted of 30 samples, unique from those used to create the predictive models. The quality of the prediction models was decided by a high correlation coefficient (RSQ), a low error of calibration (SEC), a low bias and a low error of prediction (SEP).

SEC, Bias and SEP were calculated by the following formulae:

$$\text{SEC} = \sqrt{\frac{\sum_{j=1}^{m}(yj-y\prime j)^2}{m-1-q}} \qquad \text{Bias} = \frac{\sum_{j=1}^{n}(yj-y'j)}{n} \qquad \text{SEP} = \sqrt{\frac{\sum_{j=1}^{n}(yj-y'j)^2}{n}}$$

with yj the concentration given by the primary method, y'j the concentration predicted by the NIRS for the sample j, m the number of samples in the calibration set, n the number of samples in the validation set and q the number of terms in the regression.

Table 2 lists the coefficient of correlation (RSQ), standard error of calibration (SEC) and bias for calibration models in the low to high concentration range (Calibration Set). As Table 2 shows, the predicted sucrose, betaine and RDS concentration correlate (RSQ) very strongly with those levels measured by the primary method (GC, HPLC or Refractometry).

Table 2.

Model	Constituent	Calibration Set				Validation Set	
		Mean (%m/m)	RSQ	SEC	Bias	Mean (%m/m)	SEP
T5	Sucrose	0.45	0.903	0.12	0	0.57	0.09
T4	Sucrose	21.33	0.988	0.23	0.008	22.18	0.19
T8	Sucrose	60.20	0.967	0.45	-0.007	61.02	0.38
N7	Betaine	0.65	0.999	0.02	0	0.57	0.02
N5	Betaine	17.43	0.997	0.24	-0.002	18.11	0.23
N6	Betaine	49.60	0.988	0.48	-0.002	50.26	0.34
T11	RDS	4.29	0.990	0.11	-0.001	4.07	0.09
T7	RDS	28.02	0.982	0.11	0	29.22	0.08
T1	RDS	81.34	0.992	0.30	0.001	81.75	0.25

To assess the accuracy of the NIR technique, samples independent of those used to build the models, were collected for each of the factory streams. Each, of the 30 validation samples collected, were analyzed on either the GC, HPLC or Refractometer to establish a true value. The samples were then scanned on the NIR spectrometer. The developed NIR predictive models were utilized to arrive at an estimated sucrose, betaine or RDS concentration. If a NIR prediction model is not robust, the SEP value increases with regard to the SEC value. This is mainly due to two possible causes: (1) a bias exists between the actual and predicted values of the validation sample set; (2) an additional dispersion of the values of the validation sample set after the bias has been removed. The standard error of prediction (SEP) for the validation sample set is very low and similar to the standard error of calibration (SEC). Clearly the sucrose, betaine and RDS prediction models are fit for use.

Advancements in NIRS make it a preferred measurement technique to efficiently monitor industrial process variation. The acceptance of this technique has been growing mainly because of its quick sample turnaround time, compared to conventional chemical techniques. In the sugar industry, it is likely that in the coming years, NIR spectroscopy will be a universally accepted and a preferred substitute over currently adopted methodology.

ADVANTAGES OF NIR SPECTROSCOPY

- It requires minimal or no sample preparation. There is no need for reagents or sample handling tools. This decreases analytical costs and sample handling time.

- Sample Turnaround Time (TAT) is rapid, almost instantaneous.

- The measurement technique can be designed to measure single or multi-constituents.

- NIRS is a measurement technique geared towards process control.
  The instrument is very robust, as it has no moving parts.

- NIRS measurements generally have better precision than primary analytical techniques because there is no need for sample preparation. The NIRS measurement accuracy is similar to accepted primary techniques.

- NIR spectrometers are very simple to operate and maintain once the analytical methodology has been developed.

LIMITATIONS OF NIR SPECTROSCOPY

- NIRS is not a primary analytical method. There are no available models that take into account the interaction between NIR energy and matter. Calibration is purely empirical.

- Developing accurate and robust prediction models require a significant investment in human resources. Model development requires using a large number of samples, which encompass all expected variations in the physical (matrix) and chemical properties of the sample.

- This technique is not a trace-method. It can only be applied to major sample constituents. A “Rule of Thumb” is that NIRS can measure organic constituents above 0.1%.

Summary and Conclusions:

The purpose of this work was to research the application of NIR spectroscopy for sucrose, betaine and RDS detection in molasses separator streams. Despite the complex sample matrices, seen in separator samples, the results observed show that each of the primary methods for quantifying sucrose (GC and/or HPLC), betaine (HPLC) and RDS (Refractometry), can be substituted by NIR spectroscopy for the rapid and non-destructive assessment of these constituents, in streams surrounding the coupled loop molasses desugarization separator.

This methodology has not yet been rigorously assessed in a factory setting, but when compared to current separator process control, based on RDS, apparent purity and conductivity measurements, the application of NIRS offers improvement. Due to the rapidity of the technique and the specificity of sucrose and betaine detection, this measurement technique should allow operators to make informed decisions regarding separator adjustments. This real time separator monitoring and tuning should result in higher initial sucrose and betaine yields, resulting in lower factory energy requirements and lower chemical and water usage.

Acknowledgements:

The authors would like to thank Jason L. Anderson (Twin Falls factory laboratory) and the Nampa factory laboratory personnel for their assistance with this project.

References:

1. W.F. McClure, J. Near Infrared Spectrosc. 11, 487-518 (2003).

2. Blanc, M. and Villarroya, I. NIR Spectroscopy: A Rapid-Response Analytical Tool. TrAC Trends in Analytical Chemistry, 21, 240-250 (2002).

3. Barnes RJ, Dhanog MS, Lister SJ: Standard Normal Variate Transformation and De-Trending of Near-Infrared Diffuse Reflectance Spectra. Appl Spectrosc 1989, 43(5):772-777.

Alternaria Leafspot - Control Measures and Fungicides

Dennis Bischer, Agronomist
Michigan Sugar Co.

Pioneer · Big Chief Michigan Sugar

Michigan Sugarbeet REACH Research & Education Advisory Council

Alternaria Leafspot

- History of alternaria in Michigan
- Fungicide efficacy and treatments
- Conclusions and observations
- Strategies in Michigan

Pioneer · Big Chief Michigan Sugar

Michigan Sugarbeet REACH Research & Education Advisory Council

Alternaria Leafspot in Michigan

- Low levels of disease for a long period of time
- Recent years have had significant amounts of infection
- Factors for increased disease levels
  - Susceptible variety planted on a significant percentage of acres
  - Decreased fungicide activity
  - Increased inoculum levels

PIONEER · BIG CHIEF MICHIGAN SUGAR — MICHIGAN SUGARBEET REACH Research & Education Advisory Council

Cercospora and Alternaria Leafspot

- Both diseases are present in Michigan
  - The severity of each disease is equal
- Created challenges
  - Fungicides vary in their efficacy to both diseases
    - Some fungicides are good on cercospora but poor on alternaria
  - Length of time for disease control has increased
    - Start earlier and spray later

PIONEER · BIG CHIEF MICHIGAN SUGAR — MICHIGAN SUGARBEET REACH Research & Education Advisory Council

Alternaria Leafspot

- Infects between 50-90 degrees F
- More severe in unhealthy plants
- Varieties vary in their tolerance
- Fungicides vary in their efficacy
- Economic impact is similar to Cercospora leafspot

Control of Alternaria Leafspot with Fungicides

- Trials have occurred over multiple years and multiple locations
- Evaluated the efficacy of fungicides for control of Alternaria leafspot
- The following trial is a good representation of multiple test sites

Control of Alternaria Leafspot with Fungicides

- Plot size: 6 rows (22 inch) x 38 ft
- 3 replications
- Monosem research planter 4.1 to 4.5 inch seed spacing
  - Final population of 200 beets/100 ft
- Compressed air small plot sprayer
  - 3 mph
  - 25 gallons per acre
  - 100 psi

MICHIGAN SUGARBEET REACh Research & Education Advisory Council

Control of Alternaria Leafspot with Fungicides

- Variety used-C-RR059
  - Susceptible to alternaria
  - Moderately tolerant to cercospora
- Multiple fungicides were tested for their efficacy
- The trial was NOT inoculated
- Disease pressure was severe

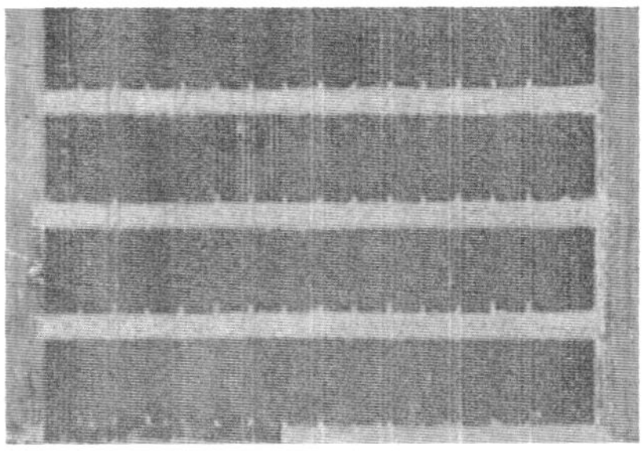

MICHIGAN SUGARBEET REACh Research & Education Advisory Council

Control of Alternaria Leafspot with Fungicides

Treatment	Tank Mix	No of Apps	Rotational Fungicide
Echo(chlorothalonil)	None	8	None
Super Tin	Topsin + Manzate	7	Badge(4 apps)
Manzate	Badge(copper)	8	None
Super Tin	Manzate	7	Badge(4 apps)
Badge(copper)	None	8	None
Manzate	None	8	None
Minerva Duo	None	7	Badge(3 apps)
Priaxor	Manzate	7	Badge(4 apps)
Headline	Manzate	7	Badge(4 apps)
Inspire XT	Manzate	7	Badge(3 apps)
Minerva	Manzate	7	Badge(3 apps)

PIONEER · BIG CHIEF MICHIGAN SUGAR

MICHIGAN SUGARBEET REACh Research & Education Advisory Council

Control of Alternaria Leafspot with Fungicides

» Treatments were rated for % leaf damage
» Treatments were harvested and tested for:
  > RWSA-sugar per acre
  > RWST-sugar per ton
  > Tons/acre

PIONEER · BIG CHIEF MICHIGAN SUGAR

MICHIGAN SUGARBEET REACh Research & Education Advisory Council

% Leaf Damage

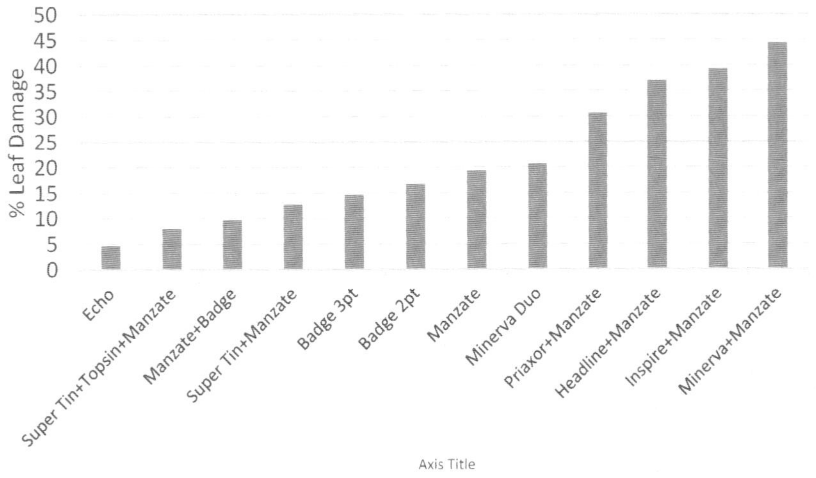

% Leaf Damage and RWSA

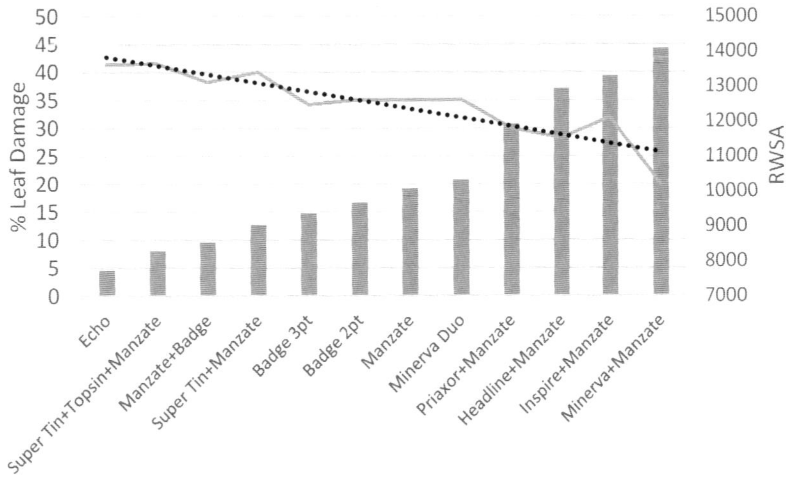

% Leaf Damage and Tons/Acre

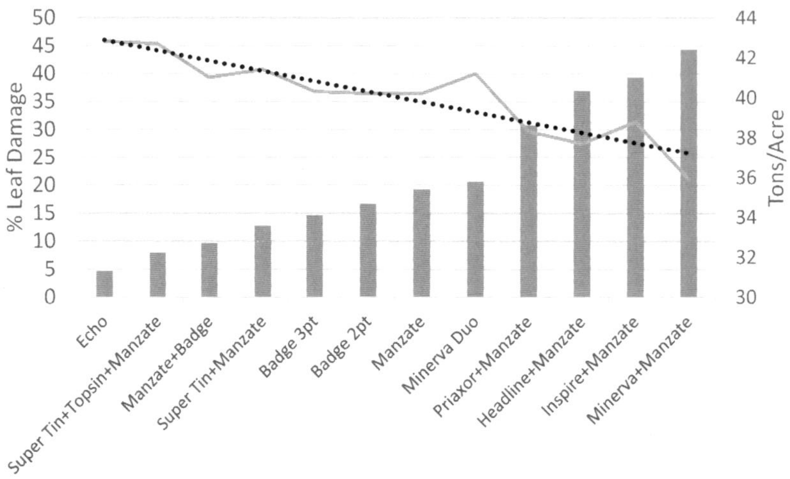

% Leaf Damage and RWST

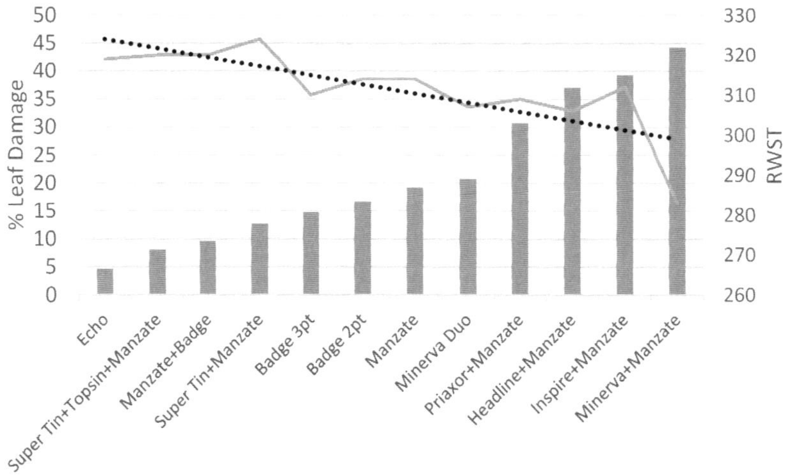

Comparison of Treatments

Super Tin + Topsin + Manzate
8.0% Leaf Damage

Priaxor + Manzate
30.7% Leaf Damage

PIONEER · BIG CHIEF MICHIGAN SUGAR

MICHIGAN SUGARBEET REACh Research & Education Advisory Council

Comparison of Treatments

Super Tin + Topsin + Manzate
8.0% Leaf Damage

Inspire XT + Manzate
39.3% Leaf Damage

PIONEER · BIG CHIEF MICHIGAN SUGAR

MICHIGAN SUGARBEET REACh Research & Education Advisory Council

Comparison of Treatments

Priaxor + Manzate
30.7% Leaf Damage

Inspire XT + Manzate
39.3% Leaf Damage

Comparison of Treatments

Super Tin + Topsin + Manzate
8.0% Leaf Damage

Untreated
100% Leaf Damage

Conclusions and Observations

» Fungicides vary in their efficacy for alternaria control

Alternaria
1. Super Tin
2. EBDC
3. Strobilurin
4. Copper
5. Triazole

Conclusions and Observations

» Fungicides vary in their efficacy for alternaria control
» Fungicide efficacy is different for alternaria and cercospora

Alternaria	Cercospora
1. Super Tin	1. Super Tin
2. EBDC	2. Triazole
3. Strobilurin	3. EBDC
4. Copper	4. Strobilurin
5. Triazole	5. Copper

Conclusions and Observations

» Fungicides vary in their efficacy for alternaria control
» Fungicide efficacy is different for alternaria and cercospora
» Tank mixing fungicides reduces leaf damage

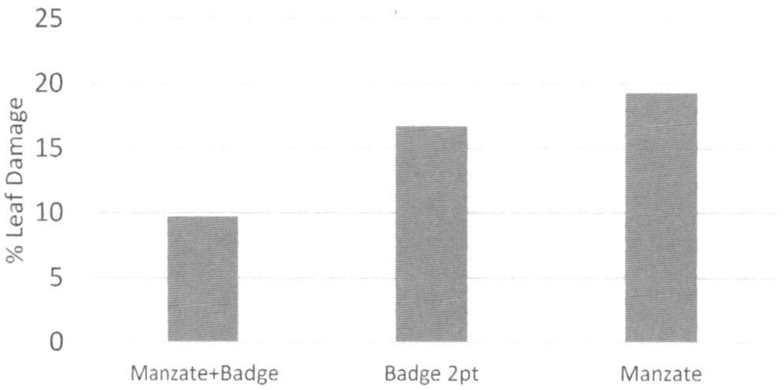

Conclusions and Observations

» The economic impact of alternaria leafspot is significant

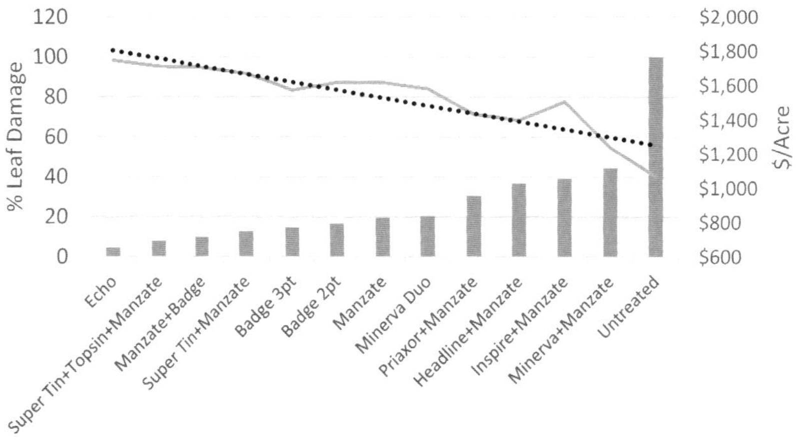

Alternaria and Cercospora Control Strategies in Michigan

» Genetic tolerance

» Timing of first fungicide

> Late June

» Application intervals of 10-14 days

» Rotate chemistry

» Tank-mix

> Fungicide applications have two or more effective products in the tank

Alternaria and Cercospora Control Strategies in Michigan

» Genetic tolerance

> Wide range of tolerance between varieties

H-9879NT
Alternaria-Good
Cercospora-Good

C-333NT
Alternaria-Fair Plus
Cercospora-Poor

C-515
Alternaria-Poor
Cercospora-Fair Minus

All varieties were sprayed 6 times with the same products and application dates

Timing of First Application

Date	Product	% Leaf Damage
		18.1
7-3	Priaxor+Manz	
7-17	Tin+Manz	
7-31	Priaxor+Manz	
8-14	Inspire+Manz	
8-30	Badge	

Date	Product	% Leaf Damage
6-28	Manzate	6.3
7-3	Priaxor+Manz	
7-17	Tin+Manz	
7-31	Priaxor+Manz	
8-14	Inspire+Manz	
8-30	Badge	

Alternaria and Cercospora Control Strategies in Michigan

» Application Interval

C-333NT
Alternaria-Fair Plus
Cercospora-Poor
6 fungicide applications
3.9% Leafspot Damage

C-333NT
Alternaria-Fair Plus
Cercospora-Poor
5 fungicide applications
41.7% Leafspot Damage

Alternaria and Cercospora Control Strategies in Michigan

» Tank-mix
  > Fungicide applications have two or more effective products in the tank

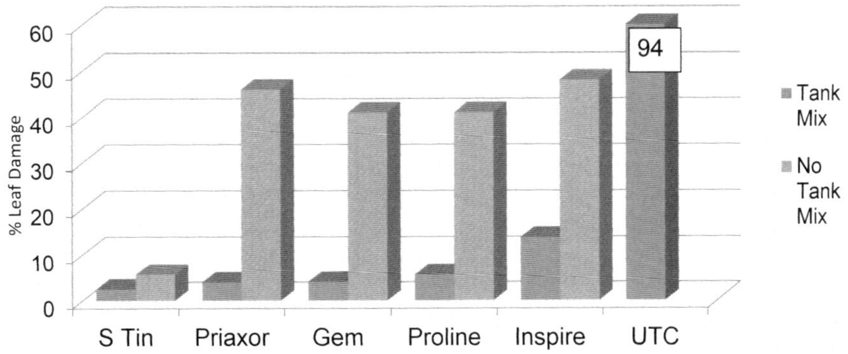

Alternaria and Cercospora Control Strategies in Michigan

» Genetic tolerance
» Timing of first fungicide
  > Late June
» Application intervals of 10-14 days
» Rotate chemistry
» Tank-mix
  > Fungicide applications have two or more effective products in the tank

Thank You!

Larson, Emma L.[1], Metzger, Michael S.[1], Peters, Thomas J.[2], and Lystad, Alexa L.[2], [1]Minn-Dak Farmers Cooperative, 7525 Red River Road, Wahpeton, ND 58075, and [2]North Dakota State University and the University of Minnesota, NDSU, Dept 7670, PO Box 6050, Fargo, ND 58108-6050. **SUGARBEET SENSITIVITY TO DICAMBA AT LOW DOSE**

Dicamba-tolerant soybean were planted at increasing frequency in close proximity to sugarbeet fields in Minnesota and North Dakota. There is currently insufficient knowledge available to advise sugarbeet growers of the effects from off-target movement of dicamba. The purpose of this experiment was to evaluate sugarbeet sensitivity and to measure for accumulation of residues in leaf and root from dicamba off-target movement to a neighboring sugarbeet field, separated by 12-, 24-, 48-rows, and beyond. Experiments were conducted at two locations in 2017 and 2018 using a randomized complete block design with up to six replications at each location. Treatments were 1/10, 1/33, and 1/100 of the labeled dicamba rate for soybean application. Results indicate that conventional soybeans are more susceptible to dicamba as compared to sugarbeets. Regardless of application rate, sugarbeet leaves were prostrate to the ground a few hours after pesticide exposure and leaf petioles exhibited varying levels of epinasty. Symptomology in younger plants (4-6 leaf stage) included fused petioles and 'trumpeting'. At the highest concentration, leaves did not recover and remained prostrate to the ground for the remainder of the growing season. New leaf growth generally resumed around six to ten days after exposure, with the new leaves having crinkled leaf margins, parallel veins, or leaf strapping. Assessments of canopy density indicated a 45% and 30% reduction when subjected to the 1/10 rate, respectively. Residue analysis of sugarbeet tissue indicated the active ingredient was present in early leaf samples but was not detected at harvest.

Pesticide residues were not detected in root samples at either sampling period. Root yield, recoverable sugar content, and purity decreased as dicamba dosage increased.

Proceedings Talk

Duane Bernhardson[1], Elke Hilscher[2], Heiko Narten[2], Stefan Meldau[2].
[1]KWS Innovations, 1500 S 40th St Grand Forks, ND 58201
[2]KWS SAAT SE, PF 1463, 37555 Einbeck.

Heterogeneity of sugarbeets- challenge for recoverable sugar analysis of grower's field samples

The most error-prone process in sugarbeet quality analysis is the preparation of brei with the subsequent analysis. Investigations of more than 50 years show that the quality components not only vary within one beet, but also between beets. In 1961 H. Lüdecke analyzed the content of sucrose, dry matter and ash in sugarbeet by separating each sugarbeet in 16 segments. The variations between the analyzed segments of a sugarbeet where considerable high, indicating a heterogeneity in the quality components.

In 2001 K. Mahn et al determined a decrease in quality components of a sugarbeet from the root to top. Furthermore, the increase of shoot material in a sugarbeet sample led to a decrease in the quality. In 1972 M. Burba analyzed 250 randomly taken samples, each consisting of 20 beets, from a regular population. To ascertain the optimum sample size for a representative determination of quality components, a statistical analysis was made. The optimal sample size for reliable detection of recoverable sugar was determined to be at least 50 sugarbeets.

During this period, the average sugar content in sugarbeet increased from 15 % to 18 % in Germany, due to progress in breeding, which may have influenced the distribution of the quality components within the beet. To study the distribution of several quality components within beets, 2 varieties (higher tonnage variety and higher sugar content variety) were investigated. From every variety, 100 uniform beets were selected and sorted by size to process homogenous groups of 5 to 7 beets. In average 16 replications per variety were produced. Every beet was divided into 6 horizontal sections. The slices were cut along specified circular vascular bundles to produce 11 segments in total (Figure 1). The corresponding segments (same segments of all beets) were pooled and processed by grinding and homogenization with ceramic knives in a mill (Grindomix).

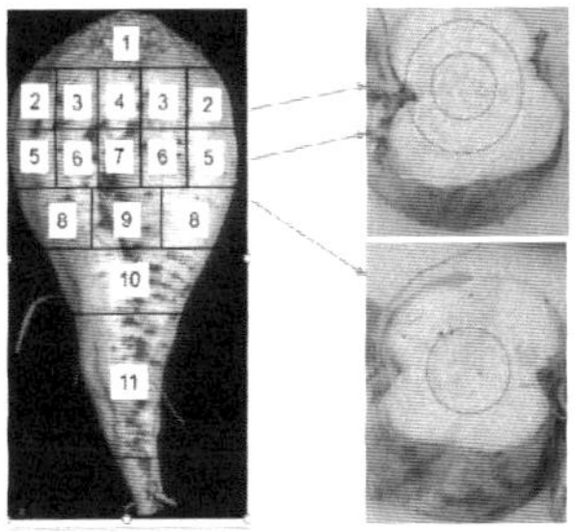

Figure 1 Definition of beet segments

The so produced brei samples were analyzed using different analytical methods to determine the quality components as Polarisation, Sodium; Potassium; Amino-

Nitrogen; soluble Nitrogen; total Nitrogen; Dry Matter; Glucose. Standard molasses loss (SML), calculated according to Buchholz et al (1995) was used for evaluating recoverable sugar in the different beet segments (Formula 1):

$$SML= 0{,}12(w' K+Na) + 0{,}24 (w'AmN) + 0{,}48 \quad (1)$$

For the evaluation of the distribution of the quality components in beet segments different analysis were performed:

- comparison of values with average beet weight,
- outlier check in raw data based on residuals of an ANOVA
- Comparison of segments with ANOVA and Tukey HSD

The results show significant differences between segments for all components. In general, the concentration of recoverable sugar increased from root tip to upper root. In the central marc the content of recoverable sugar is lower compared to the root tissue outside. The highest content of recoverable sugar was determined in the middle root tissue. The distribution of quality components within the sugarbeets was comparable in both varieties. The results indicate a higher heterogeneity for recoverable sugar in sugarbeet in comparison to previous investigations

An additional trial was performed to check for a possible decrease in beet to beet heterogeneity due to the improved field management during the past 50 years. One truckload of sugarbeet was divided in 459 individual samples. Each sample was processed and analyzed in a quality lab. The results for the recoverable sugar content show the huge heterogeneity in one growers truckload. Recoverable sugar content ranged from 11.9 % up to 16.2 % with an average of 14.5 % and standard deviation of 0.79 %. To evaluate the effect of sample size, random sample results were joined to virtually larger samples. **Error! Reference source not found.** shows the expected decrease of the standard deviation by increasing the simulated sample size. The original average sample size was 11.5 beets per sample. By merging two samples the standard deviation decreased from 0.79 % to 0.56 % and by merging four samples it decreased further to 0.39. Even when increasing the sample size to 200 beets per sample the standard deviation remains at 0.19 % and is still multiple times higher than the analytical error (<0.05 %). The results of this calculation indicates the importance of sampling and specifically choosing of the right sample size, that match to the desired accuracy limit.

Chart 1: Influence of sample size on sample heterogeneity

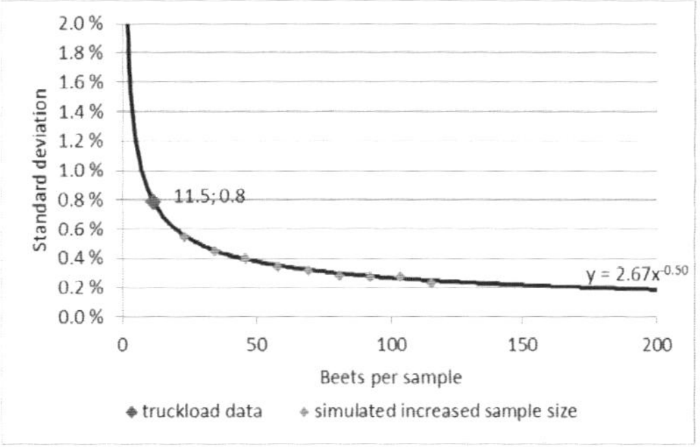

All results emphasize the importance of sampling, sample size and sample preparation for reliable quality analysis of sugarbeets.

The conventional method in sample preparation and analysis requires a central lab, a logistics system for samples and trained employees. Furthermore it causes chemical waste and is a time-consuming process. The overcome these high requirements, an innovative approach to analyze the full heterogeneity of sugarbeet in 20 seconds is presented. This technology is based on a diode array- NIRS- spectrometer, which works automatically, is mobile and can be implemented with less effort in sugar factories.

The method consists of a chopper, which chops the 20-40 kg of sugarbeet in few seconds in small, uniform pieces. A conveyor belt, equipped with a pressure roll, produce a homogenized and uniform stream of chopped beets. A NIR- spectrometer in wavelength range 850-1650 nm scans in diffuse reflection the surface of the chopped beets in distance of 200 mm (7,9 inch), every 40 ms a spectrum is scanned. The spectrometer records on single spectrum every 40 ms, leading to 400 spectra (measurements) per sample, depending from sample size. Using this approach, the whole heterogeneity of a beet sample is captured. Chemometric methods were used to analyze the spectral data and correlate it with reference data from conventional methods of sugarbeet analysis in the quality laboratories of KWS. Multivariate calibration models were developed to predict recoverable sugar (**Error! Reference source not found.**).

Table 1: NIRS calibration statistics for recoverable sugar content of sugarbeets

Trait	Reference	calibration				validation					
		N_{cal}	min	max	SD	N_{val}	min	max	SD	R	SEP
Recoverable sugar [%]	**calculated from Polarisation, Na, K and α-amino-N (ICUMSA)**	1756	9.0	20.6	1.65	8411	9.5	20.6	1.43	0.95	0.43

Additional calibrations for sugar, dry matter and marc content were also developed and can be used simultaneously. The NIRS calibration models have proven their predictive performance in worldwide breeding activities of KWS SAAT SE for 10 years.

The heterogeneity in sugarbeet has the largest impact in the estimation of the beet quality. To receive reliable and comparable analytical results, good sampling and sample preparation is crucial. Combining an optimized sample with NIRS measurements can lead to a reduced global estimation error.

The presented BEETROMETER™ provides a fully automated sample preparation and NIRS measurement system of grower's field samples.

References

H. Lüdecke Zuckerrübenbau (1961); p.20-29
K. Mahn et al Europ.J.Agronomy 17(2002); p.29-39
Burba, M. et al Z.f.d.Zuckerind. 22(1972); p.75 – 80
Buchholz, K. et al Zuckerindustrie120(1995); p.113-121

SESVANDERHAVE

HOW BIG DATA WILL CHANGE SUGAR BEET BREEDING

Hendrik Tschoep, Juan Vegas, David Boehm & Gerhard Steinruecken

"Data is the new oil"

"Machine Learning"

"Artificial Intelligence"

Big Data

"Artificial Intelligence is the new electricity"

"Digitilisation"

"Algorithms"

"Big data vs big understanding"

Breeding in the old times…

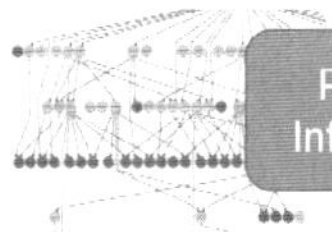

The BREEDER
"Knows and decides everything!"

Experience/
Gut Feeling

New Technologies – Drone Phenotyping

Inhomogeneities
on field trials

- Inhomogene fertilizer application
- Soil quality inhomogeneities
- Drought spots
- Spots with diseases

Correction on plot level possible using adequate statistical models

New Technologies – Drone Phenotyping

Disease Nurseries

- Cercospora beticola
- Powdery Mildew
- Uromyces beticola
- Fusarium oxysporum

High-Throughput spatio-temporal asessment of disease on plot level

New Technologies – Drone Phenotyping

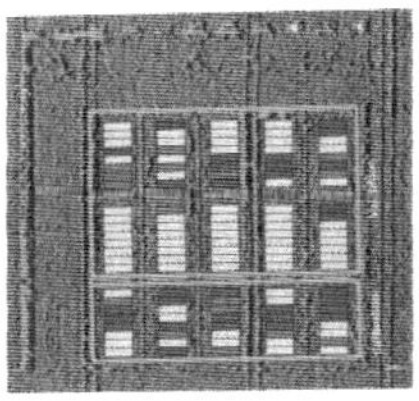

Abiotic Stress

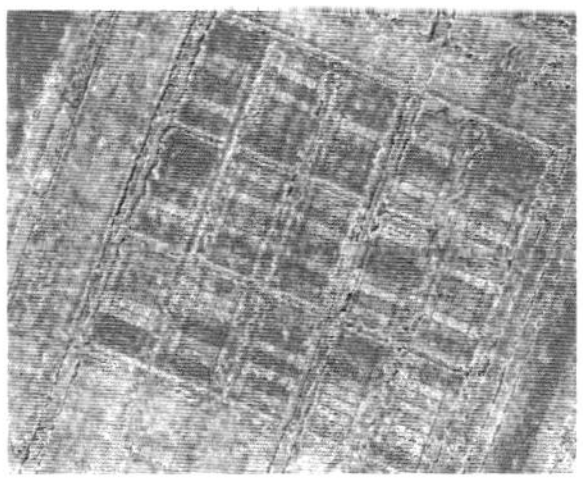

- Drought Stress
- Heat Stress

Identification of genotypes with improved stress tolerance

New Technologies – Sequencing/Genotyping

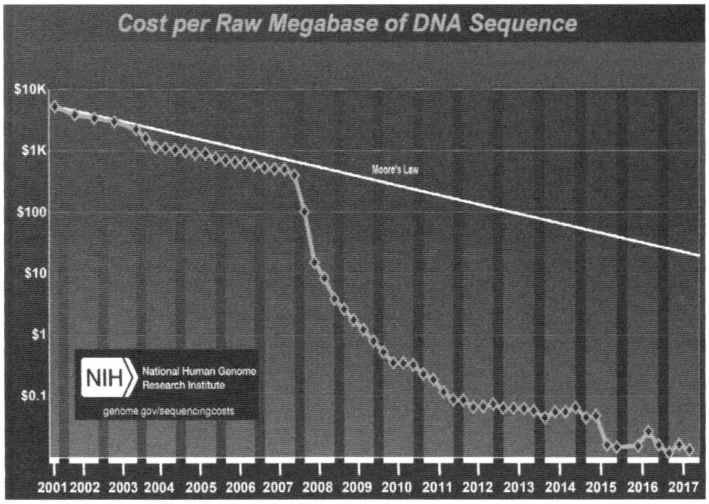

Source: NCBI: https://www.genome.gov/27541954/dna-sequencing-costs-data/

New Technologies – Genomic Selection

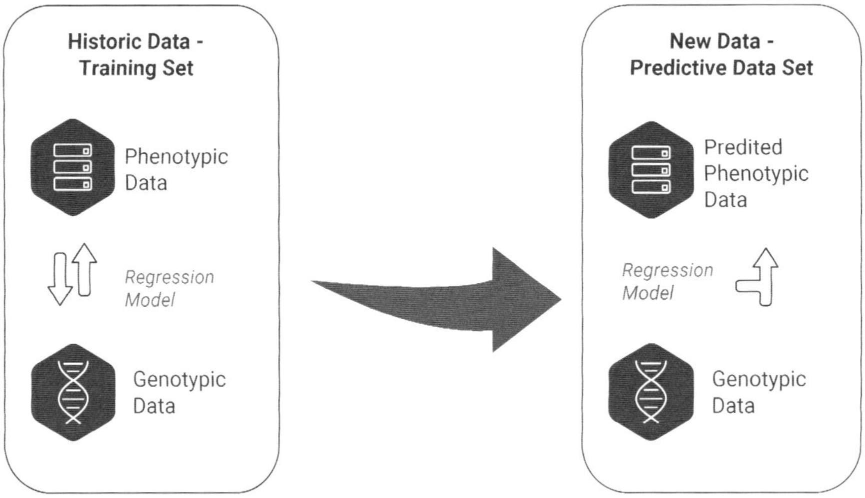

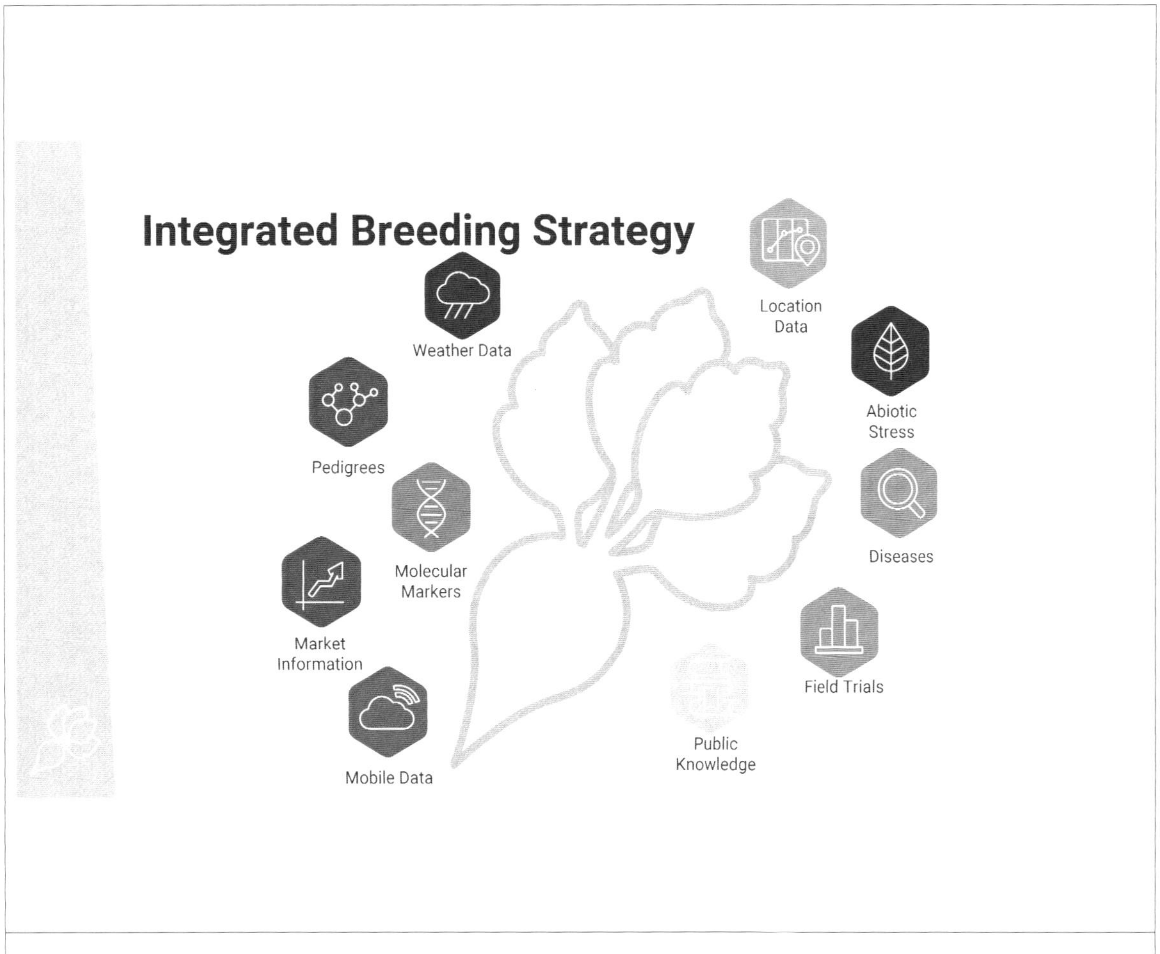
Integrated Breeding Strategy
Location Data
Weather Data
Abiotic Stress
Pedigrees
Diseases
Molecular Markers
Market Information
Field Trials
Public Knowledge
Mobile Data

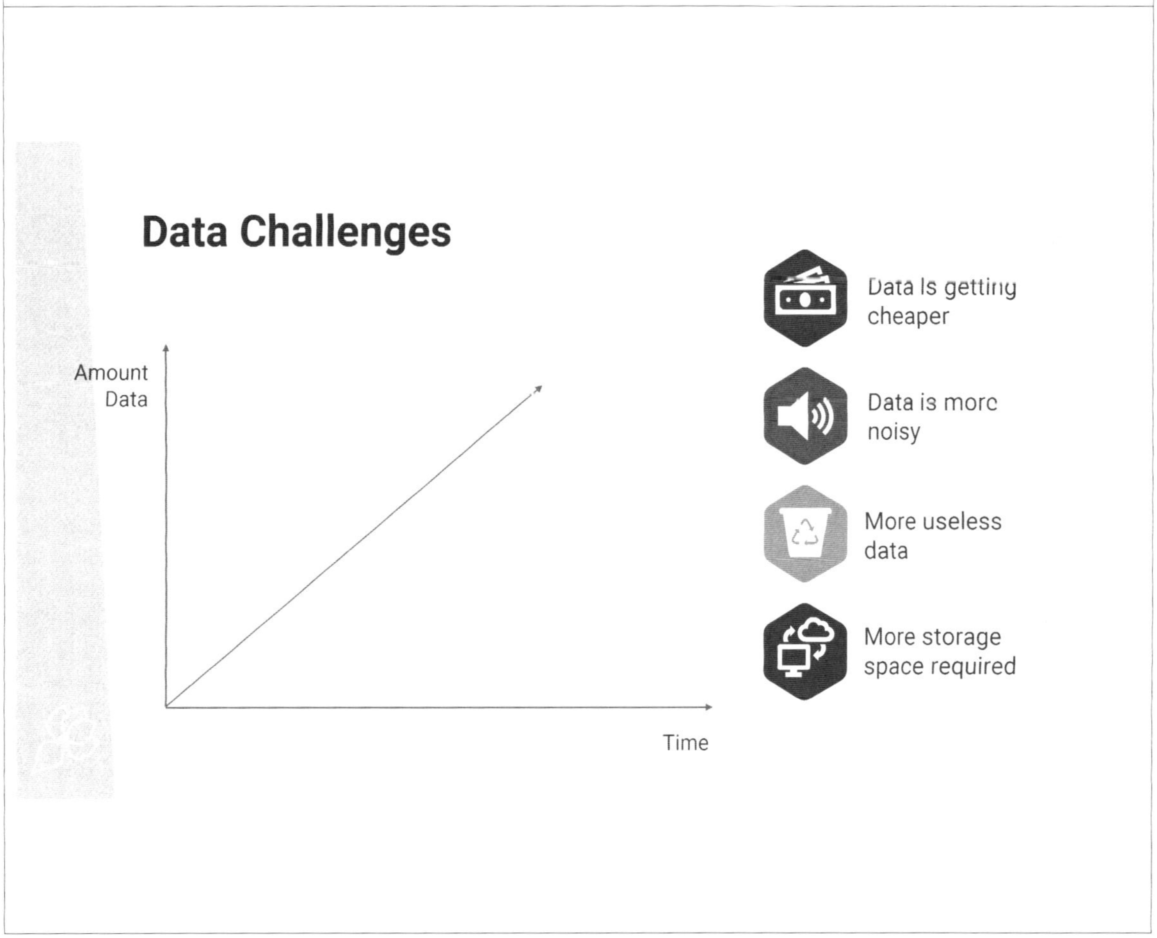
Data Challenges
Amount Data
Time
Data is getting cheaper
Data is more noisy
More useless data
More storage space required

IT Infrastructure

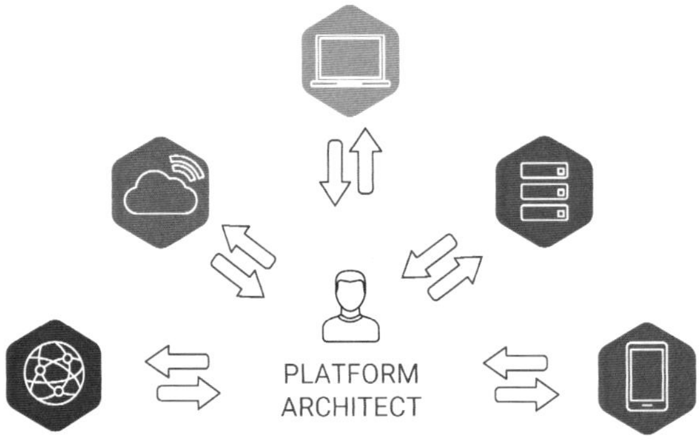

Data Analysis

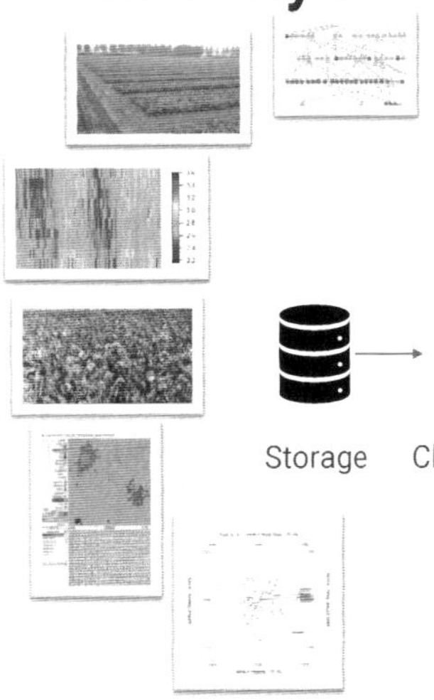

- Contant re-adjustment of the models
- Selection will be based on probabilities rather than on intuition

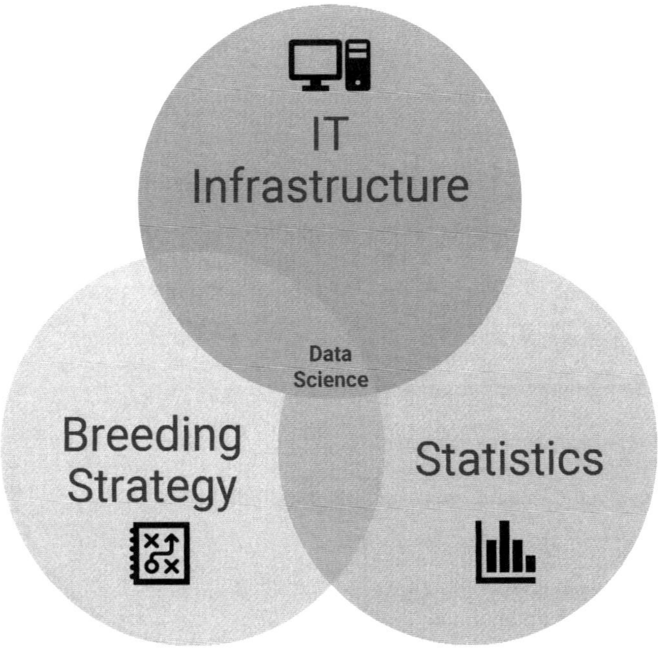

Requirement for new capabilities

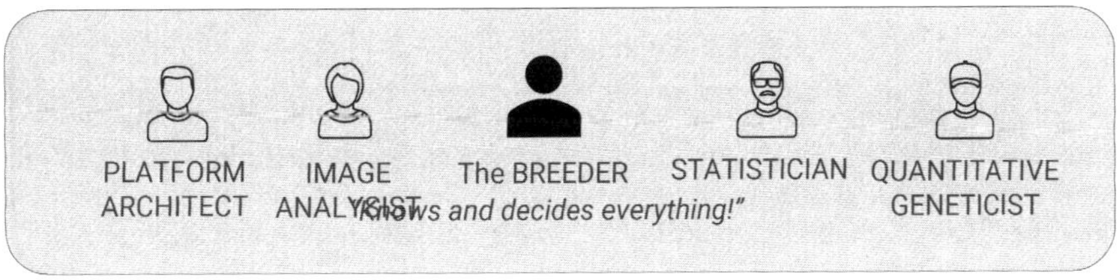

- Breeders are part of a multifunctional team
  - Modern Project Managers
- Team players with people managment skills

Big Data analysis...

- Is useless without profound analysis
- requires combined expertise from IT, statistics and breeders
- requires a team of specialists from different domains
- has to be constantly benchmarked to "on-farm performance"
- will enable the optimisation of operational activities
- will change traditional sugar beet breeding profoundly
- turns the breeder into a project and people manager
- is a challenge and an opportunity to improve and speed up variety development

SESVANDERHAVE
sugar beet seed

Hendrik Tschoep

+32 475 588 589

+32 2 16 808 246

hendrik.tschoep@sesvanderhave.com

sesvanderhave.com

Will climate change affect sugar beet crop emergence of the 21st century? Insight from a simulation study

Jay Ram Lamichhane[1], Julie Constantin[1], Jean-Noël Aubertot[1], Carolyne Dürr[2]
[1]INRA, UMR 1248 AGIR, F-31326 Castanet-Tolosan, France
[2]INRA, IRHS 1345, 42 rue George Morel, F-49071 Beaucouzé, France

Climate change will result in increased mean temperature, precipitation variability and altered frequency of extreme temperature events in many regions of the globe, including Europe. While many studies focused on evaluating the impacts of climate change on crop yields, there is less detailed information about the potential effect of climate change on crop establishment, although it is a crucial stage of the annual crop cycles. This prevents stakeholders from mobilizing adaptation strategies which may be helpful to attenuate climate change effects. Rather small adjustments (e.g. changes in varieties, sowing date and density, tillage or tactical pest management) in contrast to more systemic changes (e.g. changes in crop sequences; moving from dryland to irrigated systems or from spring to autumn sowings) may ensure successful crop establishment. Exploration of adaptation strategies to climate change using process-based models allows crop-level evaluation and adaptations of existing cropping systems.

Here we used a model-based framework, to evaluate changes in germination and emergence as well as bolting rates of sugar beet crop related to climate change in Northern France. We used the most pessimistic IPCC scenario (RCP 8.5) to generate soil temperature and water content of the seedbed using the STICS crop model (Brisson et al., 2003). We used the data obtained to feed the SIMPLE crop emergence model (Dürr et al., 2001). A total of 810 sugar beet emergence simulations were performed for a period 2020-2100, taking into account five sowing dates (mid-February, 1st March, mid-March, 1st April and mid-April) and two

First, our results provided information on the changes that will occur in future sowing conditions. The predicted cumulative rainfall for 30 days after sowing will be increasingly variable over time especially for the earlier sowings (Feb – March), with an increasing frequency of values lower than 20 mm and over 80 mm after 2050. The average maximum air temperature of the 30 days post-sowing will increase from 10°C to 12°C for the earliest sowing (mid-Feb.), with frequent values over 12°C after 2050. Second, despite the increased average temperatures, the simulated sowings indicate that the risk of reduced seed germination will remain high, with final germination rates lower than 50%, for the earlier sowing dates: 16 and 12 years out of 81 simulated for crop sown on the 15th February and 1st March, respectively. However, the use of primed seeds can alleviate this risk. The predicted bolting rates will remain very low, even for the earliest sowing dates. Overall, our results indicate that, in the future, sugar beet crop can still be sown earlier in spring than what occurs nowadays but the risk of establishment failure will remain quite high. These simulation results are useful to visualize the sowing conditions of the coming decades and pinpoint adaptive strategies for farmers or breeders. This analysis could not be possible without simulation models.

UTILIZATION OF PLANT GROWTH REGULATORS FOR SUPPRESSION OF SUGARBEET ROOT YIELD

Metzger, Michael S., Emma L. Larson and Bradley L. Schmidt
Minn-Dak Farmers Cooperative, 7525 Red River Road, Wahpeton, ND 58075

During the last decade, the sugar industry in the Red River Valley of Minnesota and North Dakota has seen a steady upward trend in overall crop yield and quality. These average annual gains are a reflection of the introduction of glyphosate-tolerant cultivars, the agronomic advancements in technology, efficiency and production practices adopted by the region's beet sugar producers. On average, commercial fields grown for Minn-Dak Farmers Cooperative from 2008 to 2017 have experienced an increase in total root yield per harvested acre of 0.98 tons per acre per year. On occasion, these root yield increases have resulted in commercial acres remaining unharvested due to the overall crop volume (tonnage) exceeding the limitations of the cooperative's storage infrastructure and factory processing capabilities. Trials conducted from 2016-2018 investigated the use of two commercially-available plant growth regulators (PGRs) as candidates to suppress late-season sugarbeet root growth without sacrificing overall sugar content or purity. Experiments were arranged in a split-plot randomized complete block design with six replications. Foliar applications of PGRs Ethophon and Atrimmec served as whole-plot treatments and were applied 30 days prior to harvest at their labeled rates of 9 fl oz/A and 2.5 pt/A, respectively. Rates ten-fold of the highest labeled rate for each product were also evaluated. Sub-plot treatments consisted of two commercially-available cultivars per experiment with a total of four cultivars (ACH 352, ACH 830, Hilleshog 4062 and Hilleshog 4302) being evaluated between the three growing seasons. Phenotypic damage (leaf speckling) from the foliar treatments of Ethophon was noted on all cultivars evaluated and a reduction in canopy height was evident across all treatments of Atrimmec. Results indicated sugarbeet tonnage could be significantly reduced utilizing the ten-fold rate of Ethophon and both the labeled and ten-fold rate of Atrimmec. Root yield suppression exceeded four tons per acre in several treatments. Even though the sugar content and purity percentage were both negatively affected by the application of these PGRs, the final sugarbeet yield and quality were such that it would still be economically feasible to harvest and process, resulting in an alternative to leaving a portion of the crop unharvested.

AUTHOR INDEX